Unraveling Reality

Behind the Veil of Existence

Ishi Nobu

"Unraveling Reality: Behind the Veil of Existence,"
by Ishi Nobu.

ISBN: 978-1-944297-26-8
Library of Congress Control Number: 2017945100

Produced in the United States of America.

First edition.

Ishi Nobu
P.O. Box 596
Roseburg, Oregon 97470
Web site: *www.ishinobu.com*
Email: *comment@ishinobu.com*

Robert D. Reed Publishers, LLC
P.O. Box 1992
Bandon, Oregon 97411
Email: *4bobreed@msn.com*
Phone: (541) 347-9882
Fax: (541) 347-9883
Web site: *www.rdpublishers.com*

Unraveling Reality is an introduction to *Spokes of the Wheel*. This book touches upon major themes in *Spokes*. In its story arc, *Spokes* explains existence: in the natural realm, in the minds of humans, and the world which humans have fashioned.

Material presented in *Spokes of the Wheel* is cumulative. Topics covered in early books are brought to further fruition later.

Unraveling Reality

Spokes 1: *The Science of Existence*

Spokes 2: *The Web of Life*

Spokes 3: *The Elements of Evolution*

Spokes 4: *The Ecology of Humans*

Spokes 5: *The Echoes of the Mind*

Spokes 6: *The Fruits of Civilization*

Spokes 7: *The Pathos of Politics*

Spokes 8: *The Hub of Being*

The Science of Existence explores the universe, and introduces the natural world. *The Web of Life* chronicles the wondrous diversity of life. *The Elements of Evolution* tells life's history, and explains how organisms adapt. *The Ecology of Humans* explores the biological interfaces of the human body. *The Echoes of the Mind* pivots on people: how they feel, think and behave. *The Fruits of Civilization* covers the consequences of human endeavors. *The Pathos of Politics* probes how polity has affected humanity. The *Spokes* series culminates in *The Hub of Being*, which is an exposition towards enlightenment.

¤✧¤

For more information about *Spokes of the Wheel*, go to: *www.ishinobu.com*. Research references for *Unraveling Reality* are at: *www.ishinobu.com/ur-notes/*.

For Sarah

Table of Contents

The Universe 1
A Matter of Energy 9
Life's Story 47
A Mental World 95
Beyond Phenomena 111
Conclusion 133
Glossary 135
People 173
Index 189

❧ The Universe ☙

One is nothing but an instrument on which the universe plays. ~ German composer Gustav Mahler

In the black of night, countless constellations of stars compose a wondrous sight. The impression the heavens made seared deep into the psyche of our ancestors, engendering myths about when and where existence began. There have been many conceptions.

3,800 years ago, in the cradle of western civilization, the Babylonians conceived a plurality of heavens and earths. A little over a millennium later, in the cradle of eastern civilization, legendary Chinese philosopher Lao Tzu had the universe originate from nothingness.

The reason why the universe is eternal is that it does not live for itself; it gives life to others as it transforms. ~ Lao Tzu

To the ancient Greeks, existence was eternal, extending over an infinity of space. But at least one Greek had a different idea. In the 6th century BCE, Anaximander of Miletus conceived a perpetual cycle of incarnation and reincarnation, powered by *apeiron*: an eternal coherence.

The primal essence of the existing objects is also the fact that when they perish, they return as dictated by necessity. ~ Anaximander

The works of Aristotle were lost during the Dark Ages. The rediscovery of Aristotle in the mid-12th century inspired many Catholic clerics, who were the keepers of scholarly knowledge in Europe at the time.

One was Robert Grosseteste, an English theologian. Contemplating God's miraculous creation, Grosseteste proposed in 1225 that the cosmos expanded from a pinpoint of light. Envisioning multiplicity evolving from an energetic singularity, Grosseteste correctly assumed that light and matter were entangled.

¤✧¤

In the 1920s, astronomers discovered that distant galaxies are moving away from us. Astrophysicists interpreted that to mean that spacetime itself is expanding.

With thermodynamics in mind, an expanding universe implied that the early cosmos had been a hot, dense, primordial fomentation. Cosmogony became the key issue: how and when the universe came to be.

> If the world has begun with a single quantum, the notions of space and time would altogether fail to have any meaning at the beginning; they would only begin to have a sensible meaning when the original quantum had been divided into a sufficient number of quanta. If this suggestion is correct, the beginning of the world happened a little before the beginning of space and time. ~ Georges Lemaître

In 1931, Roman Catholic priest Monsignor Georges Lemaître agreed with Lao Tzu and Robert Grosseteste, setting off a storm of controversy among contemporaneous cosmologists. Lemaître's radical proposal of cosmic origination (cosmogony) upset astronomers' religion.

¤✧¤

A middle-aged Albert Einstein was disturbed by the prospect of the universe starting with an explosive singularity. By 1931 he had a model of a stable cosmos, but it held a fatal flaw: the universe had to be at least 10 billion years old. Einstein found that "unacceptable."

Einstein abandoned his belief of cosmic stability as new astronomical observations indicated the universe was not as static as he had hoped. Unconvinced, English astronomer Fred Hoyle and others took up the cause of steady-state.

The term *Big Bang* was coined as a pejorative by Hoyle in a 1949 radio broadcast. Hoyle favored the ancient Greek paradigm: a steady-state cosmos, where the universe eternally existed, but continuously accreted new matter as it expanded. That there was no evidence of this worried Hoyle not a whit.

The bruited Big Bang was actually a quiet affair. No sonorous sound was made. But the misnomer does make for a catchy cosmological slogan.

Greater sophistication in considering cosmic origin suggests that existence is eternal, and comprises a multitude of universes. But this ancient conception of cyclic cosmology is now unconventional.

Based solely upon the earliest observed light from a telescope, cosmologists now surmise that the universe is 13.82 billion years old. Light only hides the darkness. We have no way to know how or when the cosmos emerged. No evidence of its origination exists.

ꕥ Galactic Expanse ꕥ

How can you look at the galaxy and not feel insignificant?
~ English filmmaker Ridley Scott

Since its inception, the cosmos has undergone enormous evolution, which is reflected in galactic dynamics.

A *galaxy* is a cluster of star systems and stellar remnants, swirling in an interstellar mixture of gas and dust. The first galaxies coalesced when the universe was only one-half billion years old.

The ballet of galaxies glide along invisible corridors. Filaments of undetectable, "dark" matter thread the universe in an invisible gravitational web. Galaxies form along these filaments. So-called *dark matter* is a euphemism for strong gravitational forces from sources which cannot be observed.

Some galaxies evolved very quickly, within a few hundred million years after the birth of the universe. There were already mature galaxies with billions of stars within the first 1.5 billion years of cosmic genesis.

There are now some 4 trillion galaxies, spread out spherically in a diameter 90 billion light years wide. Roughly half of the galaxies have light, and half are dark – detectable only by their gravitational wake. Dark galaxies have no visible stars.

Each galaxy may contain many millions or even billons of stars. Almost all visible star systems have planets.

When the universe was only a few billion years old, there were 10 times as many galaxies as there are today. Cosmic evolution reduced the number of galaxies through extensive merging. At every scale, existence gyrates in an intricate dance.

Expanding Into...?

Cosmic expansion raises an obvious question: what is the universe expanding into? The answer is: *nothing*.

Spacetime itself is delimited by the universe. The cosmos has no edge, no wrapper. All we know is that distant objects in space appear to be moving away from us in every direction, indicating an expanding universe.

The Milky Way

According to Greek mythology, the randy god Zeus had a son from a mortal woman. Zeus placed the infant, Heracles, on the breast of his goddess consort Hera breast while she was asleep, so the baby could suckle divine milk, and thus become immortal.

Hera woke up while breastfeeding Heracles. Realizing the child was not hers, she pushed him away. A jet of her milk sprayed the night sky, producing the Milky Way.

¤✧¤

Earth is tucked into an infinitesimal spot on an arm of the Milky Way galaxy, which formed 13.2 billion years ago. The Milky Way is now at least 150,000 light years in diameter, with over 400 billion stars and at least 640 billion planets, cumulatively weighing in at 3 trillion Suns.

At the center of the Milky Way is a massive black hole that barely spins. The black hole served as the gravitational pivot around which the galaxy formed. Its girth is equivalent to four million solar masses.

Despite the black hole's current sloth, that ponderous nothingness produces a fearsome whirl upon the galaxy that orbits it. The Milky Way spins at 250 kilometers per second. One revolution takes 240 million years.

The star systems at the ends of the Milky Way's galactic spirals are orbiting so fast that they should fly off. They do not because dark matter holds the galaxy together.

ꝏ Darkness ꝏ

A simple model with only 5 parameters fits all astronomical data acquired to date. Those 5 constants are: the age of the universe, the amplitude of initial fluctuations and scale dependence of this amplitude, when stars first formed, and the density of matter and atoms.

This standard model of cosmology is simple but strange. It implies almost all matter is unseen, and that most energy cannot be detected. According to known astrophysics, the cosmos should have exploded into oblivion from the get-go: an instantaneous blown bubble. Even now, there is nowhere near enough mass to hold the universe together.

Yet the universe as a whole, while dancing wildly within, is stunningly stable. Whence arose the notion of *dark matter* – an undetectable cosmic glue which holds the universe together. As the only known source of gravity is material mass, the assumed adhesive is massive, undetectable matter.

ꝏ Dark Matter ꝏ

There have been numerous hypotheses about the nature of dark matter. None have any evidence for support. Repeated searches for dark matter in various forms have found nothing to indicate that dark matter exists.

This is no minor issue. Dark matter makes up at least 96% of the mass in the universe.

> A cosmic mystery of immense proportions has deepened. The crux is that the vast majority of the mass of the universe seems to be *missing*. ~ American science reporter William J. Broad

> The word *missing* is a colloquial way of putting it. ~ American astrophysicist Neta A. Bahcall

The problem of how the cosmos is caulked together gets even more confounding. Only half of the detectable matter that supposedly exists can be accounted for.

That leaves only 2% of the mass needed to make for a stable material existence. According to physics as an accounting exercise, materiality is mostly a mirage. Once the nature of energy is taken into account, that lingering wisp dematerializes.

☙ Lightness ❧

Something is very wrong. ~ American astronomer Juna A. Kollmeier

Darkness is not the only mystery about how the universe appears to be. The amount of light in the heavens is itself inexplicable. The cosmos is presently 5 times brighter than it should be, based upon the number of light-emitting objects that have been identified. Strangely, the level of light in the early, distant part of the universe can be accounted for.

What has been surmised about the birth, evolution, and state of the cosmos cultivates more mysteries than answers. Nature presents to us a stable universe with a tidy history.

It is a mirage. If the cosmos is, indeed, a self-contained system entirely consistent with the laws of physics, most of existence is undetectable, somehow dimensionally phase-shifted away from what we may measure or experience. That means that our physics knowledge is wholly inadequate.

But that is not the end of the strangeness. All that exists has evolved in the way that it has because of what is not there. Everything is entangled with nothing.

℘ Black Holes ℘

The black holes of Nature are the most perfect macroscopic objects there are in the universe: the only elements in their construction are our concepts of space and time. ~ Indian astrophysicist Subramanyan Chandrasekhar

A *black hole* is a singularity of infinite mass and gravity. Swimming near the speed of light around a black hole are celestial objects that are either pulled in or flung out into space as part of a *quasar*, which is an interstellar light show par excellence. Anything that gets past a black hole's *event horizon* has passed the point of no return.

That an object might possess so much gravity that light could not escape it first occurred to English geologist John Michell in 1783. This idea then came to French astronomer Pierre-Simon Laplace in 1796, who produced a mathematical justification for his speculation.

German physicist Karl Schwarzschild mathematically conjured black holes in 1915; the same year Einstein introduced *general relativity*, which geometrically described space and time as a unified platform for existence. In sussing spacetime, Einstein deciphered how gravity works – as an entropic distortion of spacetime itself, rather than an active force emanating from matter, as Newtonian physics had it.

Einstein was pleasantly surprised to learn of Schwarzschild's exact solutions for general relativity's field equations. He was less pleased with black holes lurking in the background; dismissing them as merely a mathematical construct. Einstein did not think that black holes could actually form.

Following Einstein's lead, mainstream physicists disregarded all evidence of black holes for decades. Only a minority maintained that black holes were possible. It was not until the close of the 1960s that the common consensus of astrophysicists turned towards accepting the existence of black holes.

In relativistic terms, a black hole is outside spacetime. It is a nothingness without dimension, a perfectly spherical hole in the universe.

That does not mean that black holes are just infinite advertisements for immateriality, though they are that indeed. To paraphrase Lao Tzu – what is not makes what is useful.

Black holes are a crucial nothingness that arranges everything that is. The construction of the cosmos, from its early evolution, was steered by black holes. Their gravitational entropy formed the mooring around which galaxies formed.

Even now, galactic dynamics are fundamentally shaped by black holes coursing through the cosmos. A black hole crisply collapses spacetime as it goes, only to have spacetime spring back upon a black hole's departure. To say that the transition is enigmatic would be an understatement.

Black holes may form when large stars collapse after a supernova explosion. But black holes existed before there was any light, let alone stars. They are primordial. It is not understood how that could be.

The cosmos is presently peppered with black holes of all sizes. Some are no larger than a Planck pinhead; others are swollen to 50 billion times the mass of the Sun.

While the presence of black holes is doubtless, how singularities of infinity can exist is inscrutable.

> The most incomprehensible thing about the universe is that it is comprehensible. ~ Albert Einstein

⸙ A Matter of Energy ⸙

Nothing happens until something moves. ~ Albert Einstein

Physics is the study of matter in motion. A central concern of physics is *energy*: what it takes to get matter to *work*.

Work is the product of a force applied to matter. Work refers to a transfer of energy to matter. Work can be said to be energy in transit.

To get down to the nature of matter takes work. Arriving at the nature of work is a matter of energy.

But the nature of energy is, most astonishingly, nothing at all. Once you understand that, you begin to understand Nature. But the rabbit hole of reality runs deeper still.

Though existence is always in motion, that which underlies existence is motionless. Reality's nature is utter stillness. Understanding that is going to take even more work, because this conclusion collides head-on into how the world appears to be. As we are about to see, a similar paradox occurs when contrasting mundane classical physics with the arcane modern variety. There we begin.

It is a riddle, wrapped in a mystery, inside an enigma; but perhaps there is a key. ~ English politician Winston Churchill

¤✧¤

Classical physics mathematically characterized the mechanics of the readily observable. It is often called *Newtonian physics*, after its most admired decipherer, English physicist and alchemist Isaac Newton (1642 – 1727).

Newton made his mark expositing the interplay of motion and gravity. To Newton, all motion took place in the unconditional reference frame of absolute space and time.

Absolute space, in its own nature, without relation to anything external, remains always similar and immovable. Absolute, true, and mathematical time, of itself and from its own nature, flows equably without relation to anything external. ~ Isaac Newton

Classical physics concluded with *thermodynamics*: the wherefores of heat energy – both its gyral dynamics and limits. We are not naught because it's hot.

Assuming the cosmos as a self-contained system, the first law of thermodynamics is that energy is neither created nor destroyed. This conservation law is imposed by nothing more than mathematical convention. It does not characterize actuality.

> The universe does not violate the conservation of energy; rather it lies outside that law's jurisdiction. ~ Australian astrophysicist Tamara M. Davis

The other laws of thermodynamics law are concerned with *entropy*: the tendency of energy to diffuse and thereby equilibrate.

> Any other form of these laws would be so astounding as to force us to look for some more complex explanation. ~ American particle physicist Victor J. Stenger

In achieving self-sustaining integrity, life defies the laws of energy conservation. But then, biology befuddles physics. Bereft of any theory that may be applied to life, physics is diminished to describing the physical mechanics of the inorganic platform upon which organisms frolic.

In the late 19th century, physics abruptly transitioned from long-held classical mechanics to speculations about deeper truths which are hidden from our senses. Whereas the classical cast its propositions upon the ambient scale, modern physics explores the outer limits: both the tremendously tiny and the cosmologically profound.

James Clerk Maxwell intellectually grazed fields of energy, and discovered that distinct phenomena were unified underneath. Max Planck got the hots for how heat radiated, and ended up at the limits of existence. Albert Einstein wondered what the speed of light meant, and realized that everything is relative. Erwin Schrödinger transposed Planck's quantum packets into waves, whereupon existence became decidedly uncertain.

Decades after the thought experiments of these men, actual experiments confirmed that the roots of existence are stranger than can be imagined. Which meant that the consternation felt by the ancients about Nature was spot on.

Unseen Forces

Unseen forces have long awed inquisitive hominids. The power of wind and lightning, the subtleties whereby objects reliably fall to Earth, and the miracle of how life abounds, both frighten and fascinate.

Matter is easy enough to take for granted, at least until one wonders what holds it together and affords its transformative abilities. So too the Sun, until pondering what powers it, and provides for its radiation.

The *electromagnetic spectrum* is a continuum of increasing energy intensity, from longer wavelengths to shorter. High-energy wavelengths are shorter than low-energy wavelengths.

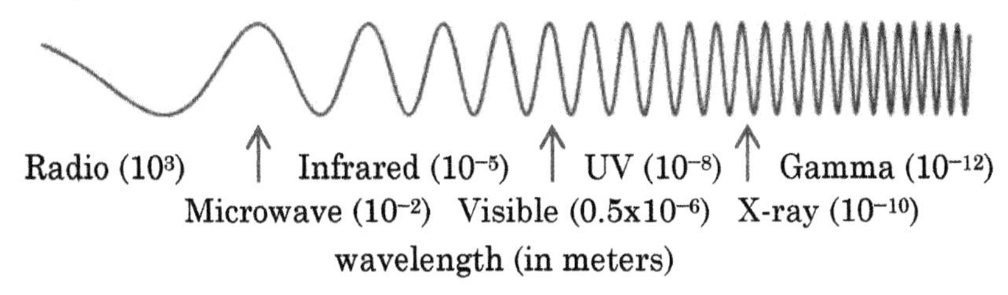

The electromagnetic spectrum exemplifies our feeble perceptual limits. Human vision – the sum total of what we see – is an infinitesimal range within a spectrum that we can comprehend only as a mathematical abstract of incredible magnitude.

The ecology of humanity for all time – all that humans can ever possibly experience – is trifling to what all of life on Earth experiences in a single day. The duration of humanity will some 300,000 years on a planet teeming with life for 6–8 billion years.* That ratio is roughly equivalent to a single breath in a person's lifetime.

---

* Life on Earth may exist for 6–8 billion years, assuming the planet and the Sun remain viable. Unless drastic steps are taken which won't be taken, humanity will have wiped itself off the planet by

¤✧¤

To understand motion is to understand Nature. ~ Italian polymath Leonardo da Vinci

As da Vinci intimated, the seminal mystery is not in what is, but in the endowment that makes movement – the motive for motion. Modern physics was forged from explaining fields and the forces that impel them; an invisible realm to us, with only effects to guide comprehension. This is the first lesson of physics: that what is apparent is either only partial or wholly deceptive.

Relativity

Nature is inexorable and immutable; she never transgresses the laws imposed upon her, or cares a whit whether her abstruse reasons and methods of operation are understandable to men. ~ Galileo

The story of relativity began in 1632, when Galileo articulated the principle of an *inertial reference frame*: that the laws of mechanics were the same wherever an observer was at constant velocity. Galileo aimed to defend Copernicus' incipient heliocentrism.

Since nothing prevents the Earth from moving, I suggest that we should now consider also whether several motions suit it, so that it can be regarded as one of the planets. For it is not the center of all the revolutions. ~ Prussian astronomer Nicolaus Copernicus

Skeptics contended that we'd feel it if the Earth were moving. Galileo refuted this with a cogent thought experiment about being enclosed in the cabin of a smoothly sailing ship at constant speed.

~2400, if not before. Humans will be lucky to last another 4 lifetimes. Civilization is a thin veneer, and very few can live without it. Even the hardiest survivors are doomed without ready fresh water and food, which is a dismal prospect under the environmental gyre of manmade pollution well underway.

> You will discover you cannot tell whether the ship is moving or standing still. ~ Galileo

With this, Galileo made the critical point of how easily appearance is confused with reality. Actuality is matter of perspective, and not to be considered truth, even when sitting upon a mountain of facts.

In 1865, Scottish physicist James Clerk Maxwell opened the door to modern physics with a field theory that unified electricity, magnetism, and light. Though all energies travel at the same speed in empty space – 299,792,458 meters per second – only light is in the visible spectrum.

This raised crucial questions. What was the medium in which these waves propagated? And light-speed relative to what?

The obvious answer, by analogy of sound through air, was that electromagnetic waves disturb an unseen medium: the *aether*. This was a revival of an ancient idea.

> There is the most translucent kind, which is called by the name of *aether*. ~ Greek philosopher and mathematician Plato

The concept of cosmic aether dated to the ancient Greeks. Aristotle proposed it as a fifth element: a divine substance comprising the heavenly spheres and bodies. The other four ancient elements were: earth, air, water and fire.

The great physics hunt of the late 19th century was to spot the all-pervasive aether, which had to be so gossamer as to have no effect on celestial bodies or feathers afloat, yet stiff enough to allow a wave to vibrate through it at unimaginable speed.

> It appeared beyond question that light must be interpreted as a vibratory process in an elastic, inert medium filling up universal space. ~ Albert Einstein

After a few decades, the search for aether came up empty-handed. Einstein felt the situation "was very depressing." Being unable to explain electromagnetism via the Newtonian "mechanical view of nature led to a fundamental dualism which in the long run was unsupportable."

The dualism to which Einstein referred was of the classic sort: a bifurcation of existence into matter on the one hand, and powerful but ethereal energy waves on the other. Einstein wanted to somehow weld the two together into an orderly unicity.

The first step down the road of relativity was coming to grips with whether the speed of light was relative or absolute. Einstein's thought experiments on how fast rays of light appeared to observers travelling near light speed left him with an incompatibility of velocities. He resolved the paradox by coming to the conclusion that the speed of light was the only sure constant.

> Light always propagates in empty space with a definite velocity that is independent of the state of motion of the emitting body. ~ Albert Einstein

If light was irreferential, something had to give. That something was time.

All reference frames must have their own relative time. Simultaneity depends upon the observer's frame of reference (*relativity of simultaneity*), as does the passage of time, which is relative to an observer's relative motion (*time dilation*). Whence Einstein's theory of special relativity.

Space too does not get away scot free under special relativity. It is subject to *length contraction*: a moving ruler that appears at rest to an observer will measure shorter than otherwise. Length contraction is only noticeable when relative motion approaches the speed of light.

In the wake of relativity, Newtonian space and time as absolutes were not only overthrown, they were reduced to an interlocked set of dimensional conduits: *spacetime*.

> Any opinion as to the form in which the energy of gravitation exists in space is of great importance, and whoever can make his opinion probable will have made an enormous stride in physical speculation. ~ James Clerk Maxwell

Special relativity occurred to Einstein in 1905. A dozen years later, he managed to apply his field theory to gravity, thereby generalizing relativity.

In doing so, Einstein proposed something indelibly bold: that gravitational fields do not just radiate through space and time. Instead, gravity defines spacetime itself.

Relativity directly states that the three dimensions of space and one of time that we experience comprise a unified 4D spacetime. But it also implies that existence has an even richer dimensionality.

> Space-time tells matter how to move and matter tells space-time how to curve. ~ American theoretical physicist John A. Wheeler

General relativity is a simple geometric theory of gravitation; a generalization of special relativity coupled to Newton's law of universal gravitation. Particles are deflected when they pass near a massive body, not because of a force per se, but because spacetime around the body is warped.

Gravity manifests as a curvature of 4D spacetime. Such distortion requires at least one extra dimension (ED) of space; an obvious implication Einstein studiously ignored, as he was uncomfortable with the idea of reality being beyond perception.

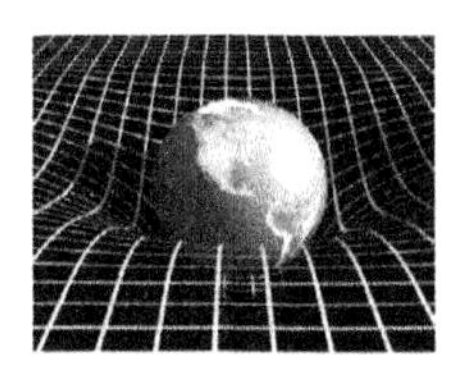

> What senses do we lack that we cannot see and cannot hear another world all around us? ~ American novelist Frank Herbert

The interrelation between mass, gravity and spacetime is an astonishing cosmic entanglement. Yet there is an even more mysterious aspect to relativity.

Extra-dimensionality demonstrates that the observable world is a fraction of existence; how large a fraction we have no way of knowing. What the measurable wavelengths of energy which our senses cannot experience demonstrate most clearly is that our direct apprehension of Nature is sorely limited.

> What we know is a drop; what we don't know is an ocean. ~ Isaac Newton

ꝏ Quantum Mechanics ꝏ

All matter originates and exists only by virtue of a force which brings the particle of an atom to vibration and holds this most minute solar system of the atom together. We must assume behind this force the existence of a conscious and intelligent mind. This mind is the matrix of all matter. ~ German physicist Max Planck

Max Planck was musically gifted. But he decided to study physics, against the advice of his physics professor, Philipp von Jolly, who told Planck in 1878: "in this field, almost everything is already discovered, and all that remains is to fill a few holes." Planck replied that he had no desire for discovery; he only wanted to understand the fundamentals.

Planck soon became fascinated with thermodynamics, whose classical laws he viewed as absolute laws of Nature. His later discovery to the contrary birthed quantum mechanics.

In 1859, German physicist Gustav Kirchhoff, who Planck later studied under, coined the term *black body* for an object which absorbs all the electromagnetic radiation which falls upon it (in other words, an utterly opaque object). He posed the inquiry: how does the intensity of the electromagnetic radiation emitted by a black body depend on the frequency of the radiation (correlated with the color of the light) and the temperature of the body?

Kirchhoff's question had been explored experimentally. But there was a serious problem with the answer that classical mechanics provided.

When a black body is heated, it emits electromagnetic waves in a broad spectrum. Experimentally, emitted black body radiation always drops off sharply on the short wavelength side. Further, as the temperature increases, peak wavelength grows shorter: visibly bluer rather than redder.

Based upon the mathematical assumption that everything is infinitely divisible, classical mechanics predicts that the energy emitted in thermal radiation is evenly distributed across all wavelengths. But that does not happen with a black body. This failure of classical thermodynamics to accurately predict the spectral characteristics of black-body radiation came to be called the "ultraviolet catastrophe."

ꟾ Quanta ꟾ

In 1894, electricity companies commissioned Planck to discover how to generate the greatest luminosity from light bulbs with the minimum energy. To get to a solution, Planck turned his attention to the black-body radiation problem.

In 1900, Planck had a theoretical answer. With great distaste, he had borrowed ideas from statistical mechanics that had been introduced earlier by Austrian physicist Ludwig Boltzmann.

Planck had a strong aversion to treating thermodynamics' laws as statistical rather than absolute gospel. Being compelled to apply statistics to get an agreeable solution he considered "an act of despair."

Planck was able to achieve concordance with experimental results via a simple formula: $E = hv$, where E is the energy of a wave, v is the frequency of the radiation, and h is a very small number that came to be known as *Planck's constant* (aka *Planck's action quantum*).

To his great consternation, what Planck found was that energy absorption and radiation was not continuous. Energetic work instead happens in discrete amounts: *quanta* of energy, with the Planck constant (h) as the quantum.* What was supposedly entirely wavelike manifested in particulate form.

> It seemed so incompatible with the traditional view of the universe provided by physics. ~ Max Planck

At first, Planck considered quantization only "a purely formal assumption" which he "did not think much about." But then he tried to stuff the quantum genie back into the bottle, and found that he couldn't.

---

* The elementary quantum of action known as *Planck's constant* is 6.626×10^{-34} joule/second in meter/kilogram/second units, with just a bit of uncertainty.

My unavailing attempts to somehow reintegrate the action quantum into classical theory extended over several years and caused me much trouble. ~ Max Planck

¤✧¤

Statistical classical mechanics requires the existence of the Planck constant, but does not define its value. Planck ushered the recognition that physical action cannot take on an arbitrary value. In other words, there is a fundamental order to Nature.

Phenomena must be a multiple of the Planck constant.* Planck's quantum of action essentially states that only certain energy levels may manifest, while values in between are forbidden to do so. Physics cannot explain why.

Existence consists of interacting fields which necessarily manifest in particulate form. Even thermal energy (heat) quantizes. We'll see that this duality is both illusory and necessary for existence.

The dynamics of quantum systems are encoded in the amplitude and phase of wave packets. ~ French quantum physicist V. Gruson *et al*

The science of the quantum world is alternately called *quantum mechanics* (accenting the statistical nature of the study), *quantum field theory* (QFT) (emphasizing that quanta are merely manifest fields), or simply *quantum theory* (which points out that all the packets under discussion are entirely theoretical, and not to be confused with fact).

ꕥ Packets of Light ꕥ

Einstein instantly appreciated Plank's discovery of quantization. He later called it "the basis of all 20th century research in physics."

Without this discovery, it would not have been possible to establish a workable theory of molecules and atoms and the

---

* Planck's constant represents the limit of empirical existence. The shortest possible distance is *Planck length*. The minimal duration is measured in *Planck time*.

energy processes that govern their transformations. Moreover, it has shattered the whole framework of classical mechanics and electrodynamics and set science a fresh task: that of finding a new conceptual basis for all of physics. ~ Albert Einstein

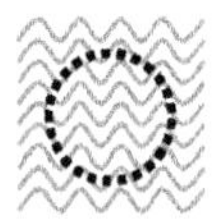

In 1905, addressing classical physics' inability to explain the photoelectric effect, Einstein argued that radiant energy consisted of quanta.

The energy of a light ray spreading out from a point source is not continuously distributed over an increasing space but consists of a finite number of energy quanta which are localized at points in space, which move without dividing, and which can only be produced and absorbed as whole units. ~ Albert Einstein*†

Echoing Plank's equation, Einstein's formula for photonic energy was: $E = hf$, where E is the energy of light at frequency f, tempered by Planck's action quantum (h).

Einstein generalized the quantum hypothesis in 1907 by using it to interpret the temperature dependence of the specific heats of solids. As a follow-on, Einstein treated thermodynamic fluctuations in two 1909 papers. Though he did not use the word *duality* or make any assertion of principle, Einstein introduced wave-particle duality into physics. This was one of several instances where Einstein failed to appreciate the implications of his assertions.

The theory of relativity has changed our view of the nature of light insofar as it does not conceive of light as a sequence of states of a hypothetical medium, but rather as something having an independent existence just like matter. ~ Albert Einstein in 1909

* The *photoelectric effect* is the glow of objects when absorbing radiation. Electrically charged particles, either electrons or ions, are emitted when a body takes on energy.

† American chemist Gilbert N. Lewis termed these particle-like packets of light *photons* in 1926.

Atomic Breakdown

Atoms were considered the most miniscule particle of matter until 1897, when English physicist J.J. Thomson found something smaller, which he called *corpuscles*. What Thomson discovered was the subatomic particle now called the *electron*.

Experimenting with cathode ray emissions, Thomson concluded that atoms were divisible into constituent corpuscles. From this he concocted a plum pudding model for atoms.

To explain the overall neutral charge of an atom, as contrasted to the corpuscle (electron) negative charge, Thomson proposed that corpuscles floated in a sea of positive charges, with electrons embedded like plums in a pudding; though Thomson's model posited rapidly moving corpuscles instead of plopped plums.

One of Thomson's pupils, English physicist and chemist Ernest Rutherford, disproved Thomas' atomic plum pudding in 1909. In its place, Rutherford imagined in 1911 a planetary atomic model: a cloud of negatively-charged electrons swirling in orbits over a compact positively-charged nucleus.

Rutherford was working with Danish physicist Niels Bohr, who conjectured in 1913 that electrons moved in specific orbits, which were regulated by Planck's quantum of action.

By 1921, Rutherford and Bohr had come up with an atomic model comprising protons, neutrons, and electrons. This model was validated in the 1950s, when atomic nuclei were manually disassembled by newly-developed subatomic particle accelerators and detectors.

Quantum Waves

In 1924, French physicist Louis de Broglie turned Einstein's quantified light inside-out by wondering whether electrons and other elementary particles exhibit wave-like behavior. A fascinated Danish physicist, Erwin Schrödinger, took the idea and ran with it. He unknowingly injected uncertainty into quantum mechanics with his 1926 publication, which described an electron as a wave function, rather than a particle at any point in time. This became known as *Schrödinger's equation.*

A most significant consequence of describing electrons as waveforms, as Schrödinger had done, was to make it mathematically impossible to state the position and momentum of an electron at any point in time. Werner Heisenberg's 1926 observation of this became known as the *uncertainty principle*: a measurement may be made to get a sense of either a quantum particle's position or momentum, but not both at the same time.

Quantum particles cannot be described as a point-like object with a well-defined velocity, because quanta inherently behave as a wave; and for a wave, momentum and position cannot both be defined accurately for any instant. This is true both practically and mathematically.

What to make of this inherent uncertainty? Physicists heatedly disagreed about what it meant.

Considering wave-particle duality a reality, Schrödinger at first took uncertainty literally. He later recanted, declaring himself utterly confused.

> I do not like it, and I am sorry I ever had anything to do with it. ~ Erwin Schrödinger

Thinking that God "does not play dice," Einstein fell back on faith to deride the uncertainty of the quantum particle/wave duality which he himself discovered.

> Quantum mechanics is very impressive. But an inner voice tells me that it is not yet the real thing. ~ Albert Einstein

¤✧¤

The particles and fields are very, very crude statistical descriptions. Those particles and those fields are not true representatives of what's really going on. ~ Dutch theoretical physicist Gerard 't Hooft, who believes the universe is a deterministic, but immaterial, information system

Louis de Broglie, who brought the matter up in the first place, came up with the *pilot wave theory*, which rendered local events determinate by virtue of a coherent force that provides every wave with its own guidance. The price was acknowledging that the entire universe was entangled – affording nonlocal interactions between particles. Though pilot wave theory is largely ignored, de Broglie was essentially correct.

Neil Bohr interpreted the uncertainty principle holistically: the universe is basically an unanalyzable whole, in which the idea of separation of particle and environment is an abstraction, except as an approximation.

Whether uncertainty is actuality remains controversial. But uncertain actuality certainly looks like the real thing.

> The statistical view is not compatible with the predictions of quantum theory. ~ English particle physicists Terry Rudolph, Matthew F. Pusey and Jonathan Barrett in 2011

Quantum theory is founded upon the premise that so-called particles are actually chunky fields. *Fields* are by definition a synchrony of waves. Denying the reality of the wave function, and its inherent uncertainty, eviscerates quantum mechanics by denying the existence of the foundation upon which the theory is built.

> The linearity of quantum mechanics is intimately connected to the strong coupling between the amplitude and phase of a quantum wave. ~ German theoretical physicist Wolfgang P. Schleich

The Standard Model

> Whether you can observe a thing or not depends on the theory which you use. It is the theory which decides what can be observed. ~ Pakistani theoretical physicist Abdus Salam

Early particle accelerators and their detectors explored the subatomic world. What they found was a multitudinous zoo. Nature's fondness for diversity became abundantly apparent at the quantum level.

Energetic collisions of protons and neutrons in atomic nuclei produced smaller, more "elemental" particles: *hadrons*. Particle accelerators proliferated hadrons into a prodigious variety, prompting Austrian theoretical physicist Wolfgang Pauli to remark:

> Had I foreseen this, I would have gone into botany.

Hadrons were found to be comprised of even tinier constituents, called *quarks*, of which there are different flavors, determined by their spin and symmetry.

Going bottoms-up – quarks combine to form families of hadrons, which join together in threesomes to form protons and neutrons (two types of baryons), which are enslaved by the nuclear force to create atomic nuclei, which combine with electrons, bound together by the electromagnetic force, to create atoms. By ionic attraction, atoms congregate into molecules, which make up everyday matter. But bear in mind that matter is nothing more than intense interactions of localized coherent energy fields, posing as something solid.

Physicists understand matter and energy in terms of kinematics and the interactions of elementary particles. *Kinematics*, which is a classical mechanics construct, characterizes the motions of bodies, without considering the forces that cause movement. This conceptual bifurcation – between matter and the forces that move matter – would live on in quantum physics' standard model.

The *Standard Model* (SM) is a myth about how existence constitutes itself from basic quantum building blocks of matter, which are constantly caressed by carriers of Nature's fundamental forces. SM proposes a set of elementary particles which compose existence at the quantum level.

The Standard Model was formulated in the 1970s. From the early 1980s, experiments verified various facets of SM. But SM cannot explain many observed quantum phenomena, nor does it include all subatomic particles discovered.

As observations have often differed from SM theory, the Standard Model has undergone repeated patchwork; so much so that SM is now a set of theorems cobbled together to render an approximate fit to what has been observed, with a lot left out.

Under the Standard Model, there are 2 elementary particle types: fermions and bosons. Whereas *fermions* are the particles that comprise matter, *bosons* are force carriers.

The SM particle zoo has 17 main characters: 12 fermions and 5 bosons, complemented by an equivalent set of antiparticles, and at least one hanger-on: the *graviton*, which is the illusive particle representing gravity.

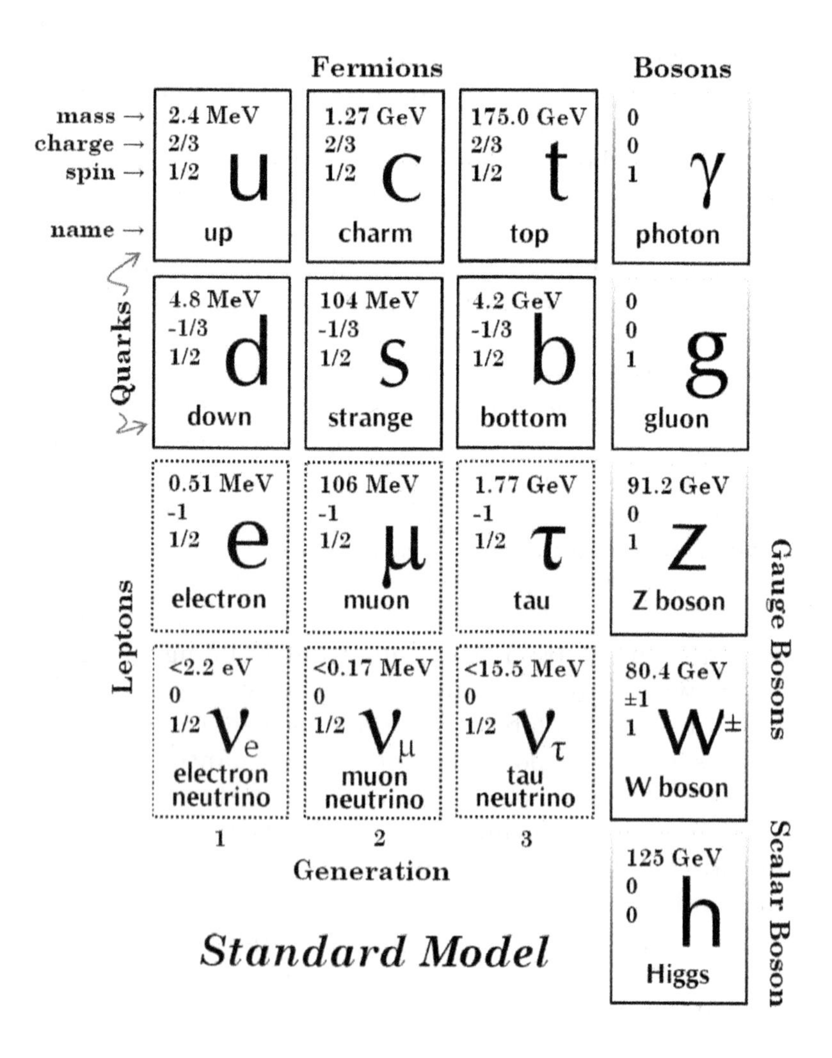

Standard Model

There is also an ancillary mob to cover various observed oddities. To conclude the parade are *virtual particles*, which are assumed to flit in and out of existence in Planck time to lend their support to the proceedings.

Photons are the best-known boson. They are the quanta of light, which manifests as electromagnetic waves. Despite not interacting amongst themselves, and remaining utterly unaffected by electrical and magnetic fields, photons somehow porter the force of electromagnetism. Except for gravity, electromagnetism is the interaction responsible for practically all phenomena encountered in everyday life.

Electrons carry the field of electricity. But they are fermions, and so it is beyond their purview to act as a fundamental force. Such a lordly task is restricted to bosons.

Hence bosonic photons, despite being decidedly standoffish, magically manage to insinuate electromagnetism everywhere, even in the dark, where no such light-hearted quanta putatively lurk. Photons finagle this fantastic feat by the same means of all quantum interactions: sheer mathematics. Equations for electromagnetism pin the proverbial tail on the photonic donkey. That, and photons are incessantly thronged by an entourage of virtual particles, which put the light touch of electromagnetism everywhere, even in the dark.

> These particles do not have a pebble-like reality but are rather the quanta of corresponding fields, just as photons are the quanta of the electromagnetic field. They are elementary excitations of a moving substratum: minuscule moving wavelets. ~ Italian theoretical physicist Carlo Rovelli

Three properties are typically used to characterize quantum particles: mass, charge, and spin. These properties are not what someone with a knowledge of classical physics would intuitively expect. Let's look at mass, which gets a lot of coverage in covering quanta.

Mass

In the everyday world, mass is considered a measure of an object at rest. Special relativity shows that rest mass and rest energy are essentially equivalent. But *rest mass* (invariant mass) does not apply to subatomic particles, as they are never at rest.

For a quantum, *mass* is a euphemism. Subatomic particle mass is a mathematical representation of its isotropy, not a measurement of anything actual.

A practical conception of *quantum mass* is the threshold energy at which a particular subatomic particle may appear; put another way, the energy required for a particular quantum species to make an appearance.

A *quantum* is not literally a particle, like some itty-bitty billiard ball. It is instead a little localized chunk of ripple in a field that puts on a particle costume. Comprehending quanta is more about wave interactions than about particle properties. Quantum mechanics is a story of coherent field behavior, not a fable of fantastically fleeting fragments.

> Particles are epiphenomena arising from fields. Unbounded fields, not bounded particles, are fundamental. ~ American theoretical physicist Art Hobson

The following is what any physicist will relate as indisputable: all matter is made of atoms formed from more elementary quanta. The actuality of quanta is that they are fundamentally coherent fields of energy.

However convincing, the physicality of quanta is nothing more than an illusory appearance. Which leads us to the crucial take-away point: that what we take for physical existence, by way of objects, is a mirage.

> Although quantum field theory tells us what we can measure, it speaks in riddles when it comes to the nature of whatever entities give rise to our observations. The theory accounts for

our observations in terms of quarks, muons, photons, and sundry quantum fields, but it does not tell us what a photon or a quantum field really is. ~ German philosopher and physicist Meinard Kuhlmann

℘ Further Afield ℘

There are many observations in the quantum world that do not fit into the Standard Model, which may be generously characterized as incomplete. Other physics theories can account for what SM cannot. But they too have flies in their ointment of exposition.

Lattice World

In 1928, English theoretical physicist Paul Dirac described his relativistic approach to characterizing a fermion field. He had in mind electrons, which have both mass and charge.

Within his mathematical solutions, Dirac found the *positron*, which is the electron's antiparticle. The positron has the same mass as the electron, but the opposite charge (positive rather than negative). Positrons were experimentally confirmed the year after the Dirac equation appeared, becoming the first antiparticle found.

In 1929, German mathematician and theoretical physicist Hermann Weyl showed that Dirac's equation could be simplified for massless fermions.

The next year, Wolfgang Pauli proposed neutrinos to explain the continuous energy spectrum coming out of radioactive decay. To respect the law of energy conservation, the neutrinos had to be chargeless.

Neutrinos were first detected in 1956. This early experimental data suggested that neutrinos lacked mass. From that, it was assumed that Dirac's neutrinos were merely massless Weyl fermions.*

---

* The masses of neutrinos are supposedly slight, but their values remain uncertain.

In 1937, Italian physicist Ettore Majorana took neutrinos to an even more ethereal state, by proposing a class of quanta that was both massless and chargeless. Majorana particles were first glimpsed in 2012.

Whereas Dirac fermions have an antiparticle counterpart, such as electrons and positrons, Majorana fermions are their own antiparticle.

Three distinct classes of fermions have been identified: Dirac (with mass and charge), Weyl (massless, charged), and Majorana (massless, chargeless). What all fermions have in common is the same *spin*, which is the direction of internal angular momentum, relative to the direction of linear momentum. Spin is the property that distinguishes fermions from bosons.

The asymmetrical spin of fermions explains why they cannot occupy the same space at the same time; but bosons can, because their spin is symmetrical. This fermionic limitation is termed the *Pauli exclusion principle*, which Wolfgang Pauli discovered in 1925.

But don't think for a Planck moment that fermions mind their manners everywhere. It all depends upon the environment they are in.

⌑✧⌑

The mathematics of existence can be quite slippery. Nature's fondness for diversity often rides roughshod over textbook behaviors. Such is the case when fermions find themselves in the tight confines of a crystalline space.

Crystals are highly ordered solids which may form any one of 230 distinct lattice structures. Figuring out the extent of lattice space groups was a tour de force of 19th-century crystallography.

In 1930, Werner Heisenberg wondered what would happen if space were quantized instead of continuous. Heisenberg was inspired to speculate about a Planck Gitterwelt (lattice world) out of a desire to rid quantum mechanics of the vexatious infinities that kept arising in equations.*

What Heisenberg got in Gitterwelt was mighty peculiar: electrons might lose their mass, or morph into protons. This strangeness drove him to abandon "this completely crazy idea."

But Gitterwelt is not unreal. An electron moving through a honeycomb lattice of graphene carbon atoms loses its mass, transforming itself from a Dirac fermion to a Weyl one. If the lattice is superconductive, the electron may drop its charge and change into a Majorana.

Lattices offer even stranger transformations. A Weyl fermion trapped in a lattice world might alter its spin to that of a boson, while still being fermionic in obeying the Pauli exclusion principle. Other quantum oddities of lattice worlds are still being explored.

The Ground State

There is no such thing as a real void. ~ Carlo Rovelli

The *ground state* is the lowest energy state of a quantum mechanical system, with supposedly zero-point energy. In quantum field theory, the ground state is called the *vacuum state*, or simply *vacuum*.

Yet the ground state is far from grounded. Vacuum energy is calculated to be 10^{113} Joules per cubic meter; unimaginably enormous power. And this is a comedown from the early universe, when the ground state was even more energetic.

According classical thermodynamics, the ground state is supposed to be at absolute zero temperature (0 K). But that is a theoretical fiction, as the ground state is not a void, or empty space. Virtual particles testify to that.

---

* Quantum mechanics' mathematics were never able to shake off infinity. So, the beautiful symmetries and inscrutable infinities that appear everywhere are purposely broken by *spontaneous symmetry breaking*, which is a statistical abuse to force equations to behave so that physicists feel comfortable with them.

Quantum mechanics teaches us that the vacuum is not empty at all, but, on small scales, contains virtual particles, their antiparticles, and the quanta of their interactions. ~ Italian particle physicist Alessandro Bettini

Virtual Particles

Our four-dimensional (4D) existence emerges from a holistic dimensionality (HD) which includes extra-dimensions (ED). Only a tiny fraction of the mass found in atomic nuclei comes from the quarks within. Over 99% comes from being bound into a hadron.

The glue that holds hadrons together are a swarming stew of *virtual particles*: subatomic dark matter that is ED, with only transient 4D appearance. Their energetic interactions multiply the mass that makes up matter.

In adding mass and other attributes to virtually all quantum bits, virtual particles from vacuum are a paradox which physicists cannot explain. The conventional comprehension is that these fleeting waveforms pop in and out of "existence" from the ground state; a rather ridiculous notion. Instead, the virtuosity of virtuality is a demonstration of dimensional phase-shifting.

The ground state is simply the limit boundary to perceptible existence (4D). Its incredible energy, and the virtual horde which incessantly emerges to add heft and vitality to phenomena, proves that existence is a chimera.

The Stability of Existence

Under the Standard Model, the Higgs boson is a grainy chunk of the *Higgs field*, which permeates all space. Elementary particles gain their mass by bathing in the ubiquitous Higgs field; a constant process called the *Higgs mechanism*.

The relationship is circular in its entanglement. While other particles gather mass by their immersion in the Higgs field, the Higgs boson depends upon those particles for its own existence. The heaviest fermion – the *top quark* – has the largest impact on the mass of the Higgs boson.

Physicists use the measurement of particle masses, and properties of the Higgs field, to deduce the stability of existence; a conclusion related to the energy of the ground state. What they conclude, according to the set of equations that define the Standard Model, is that spacetime is in a precarious predicament. The stability of the universe is at risk from a treacherous vacuum, which may at any time move to a lower energy state, instantaneously wiping out existence.

> If the Higgs mass and the top quark mass were a little bit different, we would either be in a completely stable vacuum or in an unstable vacuum that would have decayed a long time ago. The world seems to be on an edge. We don't have enough precision to say whether our vacuum is stable or not. ~ American theoretical cosmologist Sean Carroll

Ghost Fields

In the Standard Model, the masses of bosons are modified via interrelations with other bosons and fermions. This creates what are called *ghost fields*. Continually perturbing the ground state, matter radiates over these ghost fields, stirring up what has been termed *quantum foam*.

Boson-fermion interactions are called ghost fields because they are presumed to *not* exist. They are instead treated as a computational tool; a mathematical necessity to maintain consistency in the Standard Model.

But then, ghost fields are attributed as the origin of the virtual particles that appear 4D out of the ground state that comprises only vacuum energy. Virtual particles are now taken for granted as existing.

There is a paradox in granting virtual particles existence, but considering the originator of virtual particles – ghost fields – to be a fictional construct (i.e., purely mathematical mumbo-jumbo).

Ghost fields play a critical role in producing a loopy hierarchy of particles in SM, thus creating considerable complexity in the Standard Model construct. This *hierarchy problem* prompted theoretical physicists to derive a more elegant mathematical solution called supersymmetry (SUSY).

Alas, several SUSY predictions are contradicted by evidence. For example, supersymmetry predicts that electrons have a slightly oval deformation, owing to their having an electric dipole moment, which has yet to be found. Instead, electrons are perfectly spherical.

Supersymmetry is not the only alternative to the Standard Model. Another, string theory, predates SM. While not mainstream, string theory and its offshoot, brane theory, have many adherents.

❁❁❁

Before winging into strings, a brief digression on quanta that aren't, but act as if they are.

♎ Quasiparticles ♎

These particles are just smoke and mirrors, handy mathematical tricks and nothing more. Or are they? ~ English physicist Andrea Taroni

Quasiparticles are emergent phenomena that behave as quanta, but are illegitimate in the sense of being a fermion or boson per se. Quasiparticles are to quantum mechanics what epigenetics is to genetics: potent but not quite kosher. Both illustrate the deep entangled intricacy that characterizes existence.

Quasiparticles are employed to explain *oscillations*, which are the fluctuations in fields. Phonons explain mechanical vibrations. Plasmons are quantized plasma oscillations. Magnons quantize the waveform which personifies the spin property of all quanta. Excitons are the quantized excitement of nothingness (holes) that exist after electrons have departed.

String Theory

In 1907, Einstein suggested that solids came about from vibrating particles, now termed *phonons*. Einstein was guessing. The structure of atoms was not discovered until 1911.

Yet Einstein's phonon serendipity hit a note that resonated. Phonons are relevant to characterizing several exotic thermodynamic phenomena. A *phonon* is a quasiparticle that represents the excited state which brings electrons together into a Cooper pair, which synchronously perform miraculous feats like superconductivity.

> Phonons are not actually real. They are really just a way of simplifying a very complicated problem. ~ English physicist Jon Goff

Phonons were more formally conceptualized by Russian physicist Igor Tamm in 1932, as the particle form of wave-particle fields working at a particular vibration.

The first string theory was proposed in 1926, during the swirl of the quantum revolution provoked by the uncertainty principle. The idea was lost, only to be rediscovered decades later.

In 1968, Italian physicist Gabriele Veneziano was working with the *Euler Beta function*: an equation used to characterize scattering amplitude. He noticed that it could explain particle reactions involving the strong nuclear force, which binds together the nuclei of atoms. Others then realized that the equation made sense to them when they thought of subatomic particles as connected by little strings, vibrating their very little hearts out.

The concept was controversial. Shortly thereafter, the Standard Model swept aside strings as the great explainer of particle interactions. But that did not deter many physicists from continuing to pluck at string theory.

String theory postulates subatomic particles as infinitesimally thin strings vibrating through a holistic dimensionality (HD) that has more than four dimensions.

The "string" in string theory seems somewhat misleading, as the significance is that quantized fields have resonances at different frequencies, harmonically interacting with their brethren. *Vibe theory* sounds more appropriate.

In 1995, American particle theorist Edward Witten, who had been fiddling strings for over a decade, had a vision of unifying the variant quantum field theories. The result was

M-theory, which postulates 11 dimensions of spacetime: 10 of space and one of time. *'M'* stood for *membrane*.

M-theory is naturally extensible in the number of dimensions. In M-theory, a single string may be a membrane of greater dimensions.

String theorists Petr Horava and Joseph Polchinski independently extended M-strings into higher-dimensional objects: *D-branes* (a Horava term). Among other things, D-brane theory attempts to characterize string endpoints.

D-branes add rich mathematical texture to M-theory, paving the way for constructing more intricate cosmological models with greater explanatory power. Numerous *brane-world* (brane cosmology) models have emerged.

String theory has been derided by partisans for its lack of track record. But the theory has been able to explain liquidity experimentally found at trillion-kelvin conditions and near absolute zero. Meantime, the Standard Model stood mute.

The hot quark soup and ultracold lithium broth exhibited collective behavior, flowing with the lowest possible viscosity. String theory successfully modeled these phenomena as strongly coupled particles, linked by ripples traveling extra-dimensionally.

Nonlocality

> The statistical predictions of quantum mechanics are incompatible with separable predetermination. ~ John Stewart Bell

Quantum mechanics has an obvious deficiency: its mechanics. Quantifying quantum phenomena is the elephant in the room of interpreting quantum theory.

Measuring fundamental particles is an existential oxymoron. Watching a wave function collapse is a probabilistic event. The math itself is nontrivial, and the appropriateness of the bandied equations contentious.

But some quantum field phenomena have been seen. The most inexplicable is nonlocality; what Einstein called "spooky action at a distance."

Our world works on the principle of *locality*: that an object can only be affected by its immediate surroundings. In contrast, *nonlocality* is the notion that distance is ultimately an illusion.

A 1935 paper by Einstein and two other physicists posited a paradox over quantum uncertainty: that either locality or uncertainty must be true. Empirically-minded Einstein opted for locality (and certainty), thereby concluding that the wave function must be an incomplete description.

Despite upsetting the apple cart of classical physics with his relativity theories, Einstein still preferred the cosmos as comfortably commonplace. After all, relativity only applied when traveling near the speed of light, or at the scale of galactic expanse. These realms were practically abstract.

In response to Einstein's 1935 paper, Irish physicist John Stewart Bell tackled the quantum measurement problem in 1964. Whence arose *Bell's theorem.*

Science in general, and physics in particular, long assumed that both locality and objectivity were both true.

Locality means that distance affects the probability of interactions. Locality is colloquially codified in the everyday concept of *cause and effect.*

Objectivity insists that reality is ultimately objective, and therefore independent of observation. With special relativity, Einstein suggested that existence was subjective.

Bell's theorem stated that either locality or objectivity had to go. In opting for the uniformity of objective reality, Bell pitched locality.

Striving to spite special relativity, Einstein struggled to the end of his days for a theory to uphold causality; protesting against the view that there is no objective physicality other than that which is revealed through quantum mechanical formalism.

In squaring off the principle of locality against counterfactual definiteness, Bell's theorem went the other way – stating that some quantum effects travel faster than light ever can, thus violating locality.

Bell's theorem painted special relativity into a corner – rendering it applicable only at the macro scale, and irrelevant at the quantum level. Then even that corner was painted over in the 21st century, by nonlocality showing up in the ambient world.

¤✧¤

Cause-and-effect is how we understand physics in the everyday (ambient) sense. In physics, causality as predictable is termed *counterfactual definiteness* (CFD).

CFD goes to measurement repeatability: whether what has happened in the past is a statistical indicator of the future. Locality goes along with sequential causality (cause resulting in effect).

At the quantum level, CFD butts heads with locality, by stating that past probability as indicative of the future is a chimera. Instead, uncertainty always reigns.

Here we have a basic conflict. In the physics of existence, either certainty or uncertainty is true. The two are mutually exclusive.

Bell's theorem showed that quantum uncertainty was a certain reality. The principle of locality breaks down at the quantum level.

> No reasonable definition of reality could be expected to permit this. ~ Albert Einstein

Later findings demonstrated that nonlocality functions at the macroscopic level too. With spooky-action-at-a-distance a reality, superluminal (faster-than-light) effects exist. Bell's theorem of nonlocality/entanglement is considered a fundamental principle of quantum mechanics, having been supported by a substantial body of evidence.

> Nonlocality is so fundamental and so important for our worldview of quantum mechanics. ~ Swiss quantum physicist Nicolas Gisin

The supposed trade-off between locality and objective reality is a false one. While quantum locality has been disproven, there is no proof that existence is actually objective. It just appears that way as a social convention (we consider the world "objective" when others agreed with us); and so objectivity is taken axiomatically, just as locality was for so long.

> If quantum physics hasn't profoundly shocked you, you haven't understood it yet. ~ Niels Bohr

Entanglement

> That one body may act upon another at a distance through a vacuum, without the mediation of anything else, by and through which their action and force may be conveyed from one to another, is to me so great an absurdity that I believe no man who has in philosophical matters a competent faculty of thinking can ever fall into it. ~ Isaac Newton

Basic notions in physics depend upon a time continuum: cause preceding effect. The principle of locality must exist for cause and effect to work.

If causality is kicked aside, such as with simultaneous ("spooky") action at a distance, locality is violated.

Nonlocality is a well-established fact. Quantum entanglement has repeatedly been demonstrated.

The fundamental properties of chemistry rely upon entanglement. Solids form and retain their solidity via quantum entanglement of the electrons in the material. Superconductivity works through entangled electron pairs.

Superluminal communication presents a challenge to theoretical physics that has not been resolved. It is a challenge that can never be met by insisting upon the universe as a 4D closed system; an axiom of which Newton and Einstein were so confident, but simply is not so.

A practical pointer to time as an emergent property occurs by entangling particles that don't exist at the same time. In other words, nonlocality can also be nontemporal.

A scheme termed *entanglement swapping* – chaining entanglement through time between subatomic particle pairs – has been demonstrated, using four photons.

Entanglement demonstrates that time, as well as space, is *emergent*: constantly coming into being, as contrasted to preexisting and incrementally evolving, as it appears to us.

> Space and time will end up being emergent concepts; i.e. they will not be present in the fundamental formulation of the theory and will appear as approximate semiclassical notions in the macroscopic world. ~ Israeli theoretical physicist Nathan Seiberg

ꙮ Time ꙮ

> The distinction between past, present and future is only a stubbornly persistent illusion. ~ Albert Einstein

We measure existence temporally. The mind experiences time roughly through durations, scaled to our ambient sense of existence.

The present moment is a mental construct within the current 2.5 seconds; just long enough to create a comprehensible context. The experience of *now* is an ever-emergent fabrication of the mind.

> Our sense of now can be viewed as a psychological illusion based on the past and a prediction of the near future. ~ American psychologist David Melcher

Our awareness may dilate or compress time. Everyone occasionally experiences time distortion, where events unfold more slowly or quickly than strict chronology would have it.

> Time is too slow for those who wait, too swift for those who fear, too long for those who grieve, too short for those who rejoice, but for those who love, time is eternity. ~ American author Henry van Dyke

Rough sketches of experiences are stored in memory, providing a fictive cushioning of continuity. The future is a cognitive template – framing fears, hopes, and goals.

> The past is a ghost, the future a dream, and all we ever have is now. ~ American comedian Bill Cosby

In physics, time does not pass or flow. Instead, time is laid out as a map rather than a sequence. Despite relativity's twining of *spacetime*, time stands alone in its qualities.

At the macroscopic scale, time differs from space in introducing *causality*; the myth which we rely to grasp the order of events, and so ascribe their nature.

> We don't understand the fundamental quantum mechanics from which spacetime emerges. ~ American physicist Netta Engelhardt

In modern physics, *when* is always a probability, as is causality. Whereas special relativity insists that the order of events is subjective, quantum mechanics claims that uncertainty is objective.

> If quantum mechanics governs all phenomena, the order of events could be indefinite. ~ quantum physicist Fabio Costa

As such, quantum causality in the strict sense does not exist. Correlation serves as an approximation of causality in quantum mechanics. And so too in our own lives: connecting dots of coincidence and deeming them cause-and-effect.

There is a more immediate temporal issue than hanging onto the past or facing the future: how we view the world in the present moment. Thinking in terms of objects rather than processes is the greatest deception. Phenomena are nothing but localized processes, with individual dynamics spread over a wide range of duration.

There is an irony here. While the only moment is now, existence is properly perceived as a transformative continuum in spacetime.

Generally, larger the "object", the slower its process. Bosonic interactions occur at light speed, or even instantaneously (a physical impossibility which happens nonetheless). Molecular reactions typically transpire within thousandths of a second (milliseconds) to seconds. The natural lifetimes of organisms range from hours (e.g., microbes) to millennia (some trees may live for thousands of years).

Planetary objects process in geological time: millions to billions of years. Mountains may look like hellaciously solid objects, but in time they crumble. In the meantime, prodigious peaks constantly change; undergoing the same process of environmental entanglement that universally occurs with all of existence. Star systems and galaxies come and go.

¤✧¤

> Your real identity has no body and no thought. ~ Indian guru Nisargardatta Maharaj

You know you are not the same person you were a decade ago, or even yesterday, for that matter. You think of your behaviors as fluid, appropriate to the situation. When you catch yourself behaving not so appropriately, you tend to think of yourself as a work in progress.

But you don't look at other people the same way: as processes. Other people are objects, with personality traits, which you consider somewhat stable. You may wonder about another person's psychological stability, making them a flaky object, but not a fluid process (as you think yourself to be, even (especially) when flaky).

Hence, we have a natural dichotomy that is actually a built-in bias: a process-oriented view of self, but objectification of the world, including other people.

Now let's look at the other side of the coin of time.

Time stands still when awareness remains focused. This has been empirically demonstrated for localized quantum and atomic systems where observation is continuous; a phenomenon known as the *Zeno effect*.

> The quantum Zeno effect is real; a watched quantum pot never boils. ~ English astrophysicist John Gribbin

But awareness cannot cover the world at large. Whence seemingly chaotic dynamics ensue, and time's arrow takes the irreversibility with which we are so familiar.

Under the gaze of awareness there is no time. Time is a matter of inattention.

> Nothing in known physics corresponds to the passage of time. ~ English physicist Paul Davies

ꕤ Theory of Everything ꕤ

> From time immemorial, man has desired to comprehend the complexity of Nature in terms of as few elementary concepts as possible. ~ Abdus Salam

Physicists have long sought an umbrella theory that explains and connects all known physical phenomena, along with the predictive power to anticipate the outcome of any experiment within the physical realm.

Archimedes was perhaps first in describing Nature by axiomatic principles, and using them to deduce new results. The path of theoretical unification has been dreamt by many of his successors: Democritus, Newton, Laplace, Einstein, and others after him.

Some of the most recent attempts are string theory and its offshoots, though those are more particularly oriented towards an approach for producing a unified theory, rather than being a completed theory unto themselves.

The pivotal struggle has been unifying the three forces understood to operate at the quantum level (strong, weak, electromagnetism) with gravity, which is incidental at the quantum scale, but long proved tricky to bottle in equation form at that level, owing to the infinities that arise.

> We exist in a universe described by mathematics. But which math? ~ American theoretical physicist Antony Garrett Lisi

Whereas quantum mechanics has treated spacetime as unrealistically rigid, relativity reveals spacetime as mauled by gravity.

> The intrinsically quantum phenomenon of entanglement appears to be crucial for the emergence of classical spacetime geometry. ~ Canadian physicist Mark Van Raamsdonk

Mark Van Raamsdonk first proposed a way to enmesh quantum theory within relativity in 2009. The key insight involved entanglement of quantum fields, which has been observed.

Neighboring fields are typically more entangled than those farther away. This intimates that entanglement plays a crucial role in the geometry of spacetime.

> If you change the pattern of entanglement, you also change the geometry of spacetime. ~ Argentinian physicist Juan Maldacena

In the large, entanglement of quantum fields via flexible spacetime reconciles relativity with quantum mechanics. But explaining entanglement specifically entails synchronicity, which requires an instantaneous channel through spacetime to afford "spooky action at a distance." Such a shortcut is now known as a *wormhole*; a term coined by American theoretical physicist John Archibald Wheeler in 1957. The idea of wormholes is older than the word.

As a way to reconcile relativity with electromagnetic field theory, Hermann Weyl posited a wormhole theory in 1921. Einstein and American-Israeli physicist Nathan Rosen proposed wormholes in 1935, as a way to extra-dimensionally connect distinct black holes. This theoretical observation was an extension of general relativity, which renders the fabric of spacetime riddled with wormholes.

Wormholes act as the conceptual conduit of entanglement. That they exist has been repeatedly shown in experiments demonstrating quantum nonlocality.

Wormholes demonstrated by quantum nonlocality and hypothetical black-hole wormholes are the same thing, just on a vastly different scale in terms of size. This is simply scale-invariant uniformity – the physics of the universe behaving consistently at every level.

The flexible geometry of spacetime at the macroscopic level emerges from quantum entanglement, which is mathematically characterized as a tensor network. A *tensor* is geometric object that embodies a localized locus of relations. Tensor networks are used to describe various systemic entanglements, from quantum mechanics to holographic information storage to brain function.

The upshot is that entanglement defines the cosmic geometry of spacetime, as well as affording the quantum effects which allow chemistry to function. In short, existence emerges via entanglement.

☙ Matter & Energy ❧

Mass and energy are both but different manifestations of the same thing. ~ Albert Einstein

Special relativity came a corollary, captured in the succinct equation: $E = mc^2$, where mass (m) and energy (E) are different manifestations of the same thing, inexorably related to the speed of light (c).

Classical physicists conveniently multiplied an object's mass by the square of its velocity (mv^2) to come up with a useful indicator of its energy: $E = mv^2$. Einstein, in his relativistic equation $E = mc^2$, simply substituted the speed of light (c) for the classical notion of velocity (v).

The mass of a body is a measure of its energy content. ~ Albert Einstein in 1905

Despite its ostensible expression, what the formula essentially meant was that mass is a form of energy ($m = E/c^2$).

Mass-energy equivalence is astonishing in its implications. The speed of light squared is a huge number: equal to 9.5 light years. $m = E/c^2$ means that even the smallest amount of matter locks up an unimaginable amount of energy. In the mid-20th century, nuclear physicists would take this truth to a terrible fruition.

☙ The Bomb ❧

I am become death, the destroyer of worlds. ~ American theoretical physicist J. Robert Oppenheimer in 1945, inspired by Hindu mythology, on his creation of the atomic bomb

Uranium and plutonium are used for atomic weapons and nuclear power. Exuberantly exhibiting the nuclear weak force, they both decay by emitting *alpha particles*: bits of atomic nuclei.

Nuclear fuel elements are energy dense. The amount of free energy contained therein is millions of times that of chemical fuels, such as gasoline.

The massive radioactivity of nuclear weapons owes to the fact that only a tiny fraction of material in a bomb fissions. The blast strews the leftover radioactive matter hither and yon.

The American atomic bomb that destroyed Hiroshima at the close of World War II, loaded with 64 kg of enriched uranium, converted only 700 milligrams of matter into energy; equivalent to 70% of the weight of a US dollar bill. If the entire uranium package had fissioned, it would have vaporized the solar system and then some.

Despite the minute fraction of material that actually fissions in an atomic bomb explosion, the energy released is impressive, to say the least. But that spent energy is itself negligible to the energy employed keeping together the neighbors of those atoms that explosively split apart.

The power of the strong force – which holds all matter together – is unimaginable. Then consider the energy involved in black holes, which spin galaxies about.

Contrasting the strong force to the gravitational pull of black holes is an apple-and-oranges comparison. Each exudes its power in its own, unique, way.

The strong force is overwhelming at distances the size of a nucleon. Outside of that it is practically nonexistent.

Conversely, the spacetime distortion of gravity is imperceptible at the quantum scale. Gravity's tangible effect is only had at cosmological expanse.

Energies from light to dark, from subatomic to cosmic. Each with their own pattern and constraints.* Altogether of unaccountable extent, yet holistically harmonious in presenting a coherent universe.

---

* The varieties of energy illustrate Nature's fondness for diversity in all ways.

Matter

What appears as solid matter simply has no solidity. Instead, infinitesimally tiny localized fields of energy interact in specific ways to create an illusion of materiality.

Quantum interactions are intricate. They follow exquisitely detailed mathematical rules in maintaining and changing state by virtue of acquiring or shedding energy.

Particles are also very coherent in their discourse among themselves. Their masses, spins, and charges create a labyrinth which dictate how their interactions proceed. All of it is the dance of energy, ruled over by a mathematics.

The simplest atom – hydrogen – comprises a single proton, made of three quarks that follow a specific formula, and an electron, orbiting the proton at a relatively enormous distance, along a forever fluxing probabilistic path. The seething ground state, laden with virtual particles, enfolds this pair of particles, adding to the atom's heft and affecting its energetic gyrations.

Every atom is 99.999+% empty space. And this seeming void is filled with extra-dimensional energy.

Which is to say that seemingly solid matter is a very thin gruel, forgetting in the moment that matter is nothing more than coherent energy. But what exactly is energy?

Energy

We're electrical items, and when we die the energy goes somewhere else. ~ English actor Dominic Monaghan

Energy is commonly defined relatively as the ability to put matter to work. As matter is energy in a specific form, that definition is a tautology.

More formally, in physics, *energy* is: a conserved extensive property of a physical system. Parsing that definition for content leaves a void.

A *physical system* is a mathematical construct which states that a specific universe has been chosen for analysis. In other words, a physical system is an abstracted arbitrary region, where the external environment is ignored.

In physics, an *extensive property* is a quantifiable characteristic that relates to the size of the physical system.

Conserved means sustained; in this instance, self-contained. The quantity of energy within a defined system is considered constant. A physical system is presumed to be closed; that is, no possible leakage of energy.

Physicists assume that the 4D universe is a conserved physical system, with energy as its defining characteristic (extensive property). Virtual particles and dark matter, among other ED weirdness, show this critical assumption of 4D containment to be false.

Energy cannot be observed directly. It can only be calculated by its comparative state, and measured only by its effect on visible matter.

In of itself, energy has no presence. Energy is instead merely a relative reference.

What is energy? An abstraction that appears phenomenal.

> Energy is an abstract concept introduced by physicists in order to better understand how Nature operates. Since it is an abstract idea, we cannot form a concrete picture of it in our minds, and we find it very difficult to define it in simple terms.
> ~ American physicist Carlos I. Calle

> **Energy composes existence. Energy is immaterial. Ergo, existence is an immaterial composition.**

The stage upon which the universe appears is spacetime, which is defined by energy and its entropic counterpart: gravity. As energy is really nothing at all, and gravity culminates into singularities of nothingness (black holes), existence itself must be illusory. So, what is its source?

Before answering, let's first look at life, and how we think we know the world.

☙ Life's Story ❧

> However many ways there may be of being alive, it is certain that there are vastly more ways of being dead. ~ English evolutionary biologist Richard Dawkins

Biology defies physics. Science has no accounting for how life can exist. Quite simply, we accept what cannot be denied.

Vitalism

> It is not necessary to ask whether soul and body are one, nor generally whether the matter of each thing and that of which it is the matter are one. For even if one and being are spoken of in several ways, what is properly so spoken of is the actuality. ~ Aristotle

The principle of *vitalism* states that living organisms possess a fundamental force which distinguishes them from inanimate matter; what is sometimes called a *soul*.

The idea of vitalism is prehistoric. Ancient Egyptians wrote of vitalism. Second-century Greek anatomist Galen, the most accomplished medical researcher in antiquity, held that vital spirits were necessary for life.

In embracing the concept of souls, vitalism underlies most religions. A quixotic inquiry in Christian theology has been pinpointing the superiority of humans – elucidating qualities that make man unique, such as being the sole possessor of a soul, thus justifying man's dominion over the Earth.

From the mid-17th century, during the Age of Enlightenment, vitalism lost its vitality. The idea of a spark of life was abandoned; supplanted by a pseudo-scientific, religious belief in reductionist chemistry, wherein life was merely a peculiar molecular combination.

> The chemistry of life is distinguished by being both highly ordered and far from thermodynamic equilibrium. ~ Dutch physicochemist Rogier Braakman & American physicochemist D. Eric Smith

For well over a century, ambitious chemists have repeatedly tried to reproduce life's origination. All they ever got was organic muck.

¤✧¤

Not all scientists have subscribed to the absurdity that life is mere mechanism.

> The organizing principle, which according to an eternal law creates the different essential organs of the body, and animates them, is not itself seated in one particular organ. ~ Johannes Peter Müller

In the mid-19th century, German physiologist Johannes Peter Müller integrated chemistry, microscopic investigation and comparative anatomy to study physiology. Müller's 1840 magnum opus – *Elements of Physiology* – explained in detail bodily mechanics. The book had an overarching theme: the presence of a soul renders each organism vibrantly whole. Müller argued that perception of sound and light demonstrated that living organisms held a life energy for which physical laws could never fully account.

19th-century French chemist and microbiologist Louis Pasteur expressed that living organisms possessed "irreducibly vital phenomena."

Vitalism is despised by materialists, who refuse to acknowledge that order in Nature demonstrates a coherent force.

¤✧¤

Every cell in the human body repairs about a million DNA errors a day. Within one year, 98% of the atoms in an animal body or plant are replaced.

The positive energy that characterizes life is neither random nor automatic. There must be intelligent interaction in all of the atomic shuffling and molecular maintenance that bodies require. As tardigrades illustrate, this intelligence cannot be based in matter.

♎ Tardigrades ♎

Tardigrades are tiny, nearly translucent aquatic animals, found all over the world. Tardigrades are commonly known as *water bears*.

Though tardigrades arose over 600 million years ago, well before arthropods, they have a physiology similar to larger animals that evolved much later. Water bears have muscles, a complete digestive system, a brain and nervous system.

Tardigrades are tough: able to withstand extreme heat (125 °C), freezing (–272 °C), desiccation, and even survive the rigors of space. They do so via miraculous adaptations.

In preparation for drying out, a water bear curls into a ball, termed a *tun*. Then a tardigrade preserves itself as a dry husk; nothing more than tiny pearl of powder.

> When a typical cell dries out, its membranes rupture and leak, and its proteins unfold and aggregate together, making them useless. DNA will also start to fragment the longer it is dry. Since water bears can survive drying, they must have tricks for preventing or fixing the damage that cells like ours would die from. ~ American biologist Thomas Boothby

Tardigrades rehydrate and return to activity within a few minutes to a few hours. A tun may be dormant for decades and come back to life.

Tardigrades have a unique genetic ability to create proteins which preserve cells during desiccation. These proteins encapsulate the molecular components of cells in glasslike matrices to preserve them, keeping them intact.

Adding water melts the preservative proteins, allowing cellular revitalization. But what kept a tardigrade alive in the meantime?

❀❀❀

Biologists have documented countless instances of life adapting to resource opportunities and changing environmental conditions. Adaptation indisputably has an aim.

Evolutionary theorists studiously ignore this obvious teleology (goal orientation) by using the vacuous phrase "natural selection" to explain evolution. Such facile deflection points out the central problem of science based solely on factual evidence: empiricism can offer no explanation for how life, consciousness, and adaptive evolution are possible.

> Nature is not embarrassed by difficulties of analysis. ~ French engineer and physicist Augustin Fresnel

꧁ Earth ꧂

> Those who dwell among the beauties and mysteries of the Earth are never alone or weary of life. ~ American marine biologist Rachel Carson

Earth is tucked into an infinitesimal spot on an arm of the spiral Milky Way galaxy. The Milky Way has 100–300 billion trillion stars, spread out spherically with a diameter of 90 billion light years.

Earth took shape 4.55 billion years ago (BYA); one of four such rocky planets in the solar system. These hungry orbs scavenged the meager leftovers that the four gas giants had not gobbled up.

In gaining its massive girth at birth, the behemoth planet Jupiter consumed two-thirds of the leavings after the fireball Sun had gobbled 99.8% of the matter in the system.

In the eon of planetary formation, Jupiter was kind to Earth; protecting the nascent watery planet from a worse bombardment than it otherwise would have got. This fortuitous circumstance helped Earth become a habitable world.

> The intimate relationship between the vital phenomena with chemistry and its laws makes the idea of spontaneous generation conceivable. ~ English naturalist Charles Darwin

Life burst into being on Earth as soon as it could, some 4.1 BYA. Organisms proliferated into every possible niche where metabolic energy could be gotten, whether by molecular manipulation or by radiation intake.

We do not know where life originated on Earth, nor how. But it happened so quickly and fecundly that it could not possibly have been a random occurrence.

℘ Cells ☙

The atom of life is a cell. Like atoms, cells consist of a gyre of smaller functional constituents. Even the simplest cell is a dynamic system of astonishing intricacy, with specialized structures and functions. This even applies to prokaryotes, the earliest living cells.

Each cell has an outer layer which encapsulates contents and provides an interface to the outside world. Many cells need to move. Thus, for thrust, they have tails (flagella), or a herd of little feet (cilia).

Inside are the means to produce the energy needed to sustain the cell, fabricate the parts which maintain the cell, and keep the blueprints (DNA) for component manufacture and cellular reproduction.

Within each cell are vibrant networks of activity, with specific pathways for material transport and communication links for sharing information. The life of every cell, whether on its own or as part of a larger body, is an incessant exercise in intelligence.

Moreover, the active molecules and subsystems within a cell must also know what they are doing. As much can easily go wrong, intracellular enterprise is itself an astonishing orchestration.

℘ Proteins ☙

> To a large extent, the structure, behavior and unique qualities of each living being are a consequence of the proteins they contain. ~ American molecular biologists Kathleen Park Talaro & Barry Chess

Cellular components are the handiwork of proteins. A *protein* is a large, complex, organic (carbon-based) molecule (macromolecule).

Proteins are manufactured in an intracellular factory called a *ribosome*. The protein production process is intricate. To ensure properly functioning products, ribosomal machinery is itself subject to quality control.

> Because ribosome assembly is so energy costly and only desired when cells are growing, ribosome assembly is highly regulated in response to available nutrients and external stimuli.
> A quality control function exists that uses the system to do a test run. If ribosomal subunits don't pass, there are mechanisms to discard them. It's the most elegant and efficient way to produce perfect ribosomes. ~ American molecular biologist Katrin Karbstein

The basic recipes that ribosomes use to produce proteins are termed *genes*. But the actual manufacture of proteins often involves appended notes not found in genes. These are *epigenetic* modifications, which are expressed chemically in a variety of ways. The prefix *epi* is Greek for "outside of."

Proteins are a cell's population of workers, who not only maintain the cell, but also create its successor.

Once off the production line, proteins preen themselves by folding into a globe or fiber. This posturing is a product of energetic economy.

The elite, most energetic proteins are *enzymes*. Enzymes distinguish themselves by their hyperactivity in being able to catalyze and regulate biological reactions.

Within a cell, there are around 3 million proteins per cubic micron (μm), which is the smallest prokaryotic cell size. Prokaryotes are ~1–5 μm in size; eukaryotes ~10–100 μm. Proteins comprise quite a cellular workforce.

> The biological properties of a protein depend on its interaction with other molecules. ~ English cellular biologist John Wilson

The various structural facets of a protein – chemistry, shape, fold pattern, and assembly – are all significant in its functioning. In chemically communicating, physically morphing, and actively manipulating, a protein at work is an incredible sight.

Most proteins are multifunctional. But they often need to be focused to a specific task, else they might create cellular havoc, creating disease.

This is not to say that proteins are mindless molecules. Quite the contrary.

Besides knowing what to do, proteins know exactly what they are doing, and so are prepared for what needs to be done next. For example, by binding to an ion, a protein activates its work toolkit. In such a way, the state of a protein at any instant embodies a memory of its past.

Unsurprisingly, the massive armies of working proteins in a cell must be supervised, so as to properly play their roles. There are various physical mechanisms to achieve this necessary regulation.

What can never be seen under a microscope is the active knowledge needed to manage a cell, beginning with the vast workforce of proteins within. Physically speaking, a cell's life is an ongoing miracle, as the force behind the mechanics remains empirically enigmatic.

For example, the human body synthesizes ~100,000 different proteins. For cells (and bodies) to function, specific proteins must be produced in appropriate numbers in a timely manner. That takes large-scale communication and coordination. Whence forth such cognitive skill?

Once minted, neophyte protein laborers must head off to work, doing a proper job at the right location. Though tasks may be completed, the work of proteins is never done. Cellular construction needs are ceaseless.

¤✧¤

All this is just within a single cell. The different cells within multicellular organisms must continually communicate and coordinate with each other to properly function within their assigned organ (or other systematic arrangement).

Further, some host cells in a macroscopic body constantly interface with foreigners, such as resident microbes – the microbiome. That essentially means speaking a foreign language so as to exchange the molecular materials each party needs to keep going.

ꟿ Microbes ꟿ

The little things are infinitely the most important. ~ Irish-Scots writer and physician Arthur Conan Doyle

For nearly 3 billion years, life on Earth kept to tiny single cells. These productive inhabitants were prokaryotes, of two main varieties: bacteria and archaea.

In their initial eon, archaea dabbled in most every lifestyle, from eating photons to consuming chemicals, both inorganic.* Various archaea were autotrophic, heterotrophic, or saprotrophic.

Meanwhile, bacteria made a living chewing rocks; unlocking energy from sulfur, nitrogen, iron, and hydrogen. This created sediment that acted as a geological viscous lubricant, which was instrumental in generating tectonic plate subduction. By this, bacteria facilitated the rise of continents.

Cyanobacteria arose by deriving energy from fermentation, which does not require oxygen. By 3.5 BYA, cyanobacteria had acquired the quantum trick of photosynthesis: microscopic reactors capturing fleeting photons to convert carbon dioxide and water into the universal organic currency for energy: the sugary molecule ATP.

During the first half of Earth's history, the majority of life forms were probably capable of photosynthesis. ~ Columbian life scientist Tanai Cardona

Thus chlorophyll was born. It literally changed the world. Animal life on Earth was made possible by the byproduct of photosynthetic bacteria: oxygen.

At first, oxygen was poisonous to life. But organisms adapted and evolved to appreciate the accessible energy that atmospheric oxygen offered.

---

* The phototropism of archaea was much simpler and direct than photosynthesis, which is a complex quantum process.

Oxygenating the atmosphere altered every form of life. Even prodigal oxygen producers had to adapt to their own success.

¤✧¤

Microbes perfected metabolism near the optimality afforded by physical chemistry, with the slight trade-off of being able to adjust to alternative nutritional conditions. This efficiency goes a long way in explaining the diversity and staying power of microbes.

Viruses descended from bacteria ~4 BYA, slimming down to live vampirishly. But they kept their wits about them. Viral exuberance for communion with others was positively infectious.

꧁ Viruses ꧂

No phone, no pool, no pets... king of the road. ~ American musician Roger Miller in the song "King of the Road" (1964)

Viruses travel light. While hardy enough to survive the elements, they enjoy the comfort of being indoors.

Physically, a virus is little more than a gene package inside a protein coat. It cannot eat or reproduce. What a virus can do is hijack a host cell and run it to make copies of itself. To spread, a virus must then find a new host.

Viruses are everywhere; inflicting themselves on all other life. Every organism is constantly interacting with viruses.

In selecting only those sequences that may prove beneficial, the rapid evolution of viruses shows that they are intelligent genetic mavens.

Given their line of work, viruses are extensive travelers, to put it mildly. In constantly encountering one another, viruses constitute a worldwide community.

The descent of viruses has been obscured by their willful genetic maneuvers, which may result in dramatic transformations. But their importance has been indelible.

Universal Common Ancestor

Viruses are embedded in the fabric of life. ~ Argentinian biologist Gustavo Caetano-Anollés

In the 1740s, French natural philosopher Pierre Louis Maupertuis suggested that all of life had a common ancestor. The idea became something of a holy grail for evolutionary biologists: to find the organism from which all others descended. It turns out the culprit was a carrier, not a progenitor.

Archaea and bacteria are the earliest known life forms. They seem to have originated independently; but both ended up with DNA as their genetic coding regime.

In a world when rough-and-ready RNA made life less robust than it could be, viruses spread an innovation that gained universal acceptance by virtue of its durability: DNA. A critical tweak in a sugar molecule improved fidelity, and gave much better ability to withstand harsh environments.

Only viruses had the means, motive, and opportunity to unify life at the genetic level. While some viruses decided to continue to use RNA for their own convenience, evangelizing proponents of DNA inspired prokaryotes worldwide to adopt a superior solution to managing their library of life, known as a *genome*.

❀❀❀

As a regular work practice, viruses insert genes into their hosts. By this and their unwanted intrusions, viruses drive evolution.

8% of human DNA derived from viruses. In many instances, these viral contributions are critical.

Viruses are gregarious, establishing networks of connections between compatriots. Cooperation during infection is common, as the process is seldom easy.

The advantage of viral cooperation comes in taking advantage of specialized skill sets. Some viruses are better at certain tasks than others.

Decisions need to be made. For example, to boost total viral production, host cells may be granted greater longevity.

If a virus is co-infecting with a stranger instead of friends, it considers this competition. The virus will work its host cell to death as quickly as possible, to thwart its rival.

Viruses understand the difference between strategy and tactics. With their sharp minds, viruses amply illustrate that physicality has nothing to do with wiles.

♎ Influenza ♎

Influenza viruses are shrewd. Many originate in birds; hence the commonly bandied term *avian flu*. This host platform is especially convenient, as birds are social and often travel widely.

From their avian base, flu viruses are able to jump to various mammals: rodents, cats, dogs, pigs, ferrets, camels, cetaceans, and primates. These viruses are well-adapted to survive on surfaces that provide access. At every step, flu viruses are ready with an array of potential modifications they may apply to gain entry to a new host species.

Viruses also have numerous countermeasures they may employ against immune systems, which vary somewhat between species. This is why flu viruses are so successful, and influenza never defeated.

♎ Funky Con ♎

The cucumber mosaic virus infects garden-variety vegetable plants. Upon doing so, the virus makes the smell of those plants more alluring to aphids, which the virus employs as transport.

The infected plants do not live up to their odor. A single taste disgusts the aphid, which moves on.

That is a good thing for the virus, which hitched its ride when the aphid landed and sampled. If the aphid had stayed to dine, working its needle-like mouthparts deep into the plant, the virus would likely have been wiped off. Thus, the mosaic virus cleverly manages both the come-on and the brush-off to suit its needs.

The virus goes one further. It programs the plant, which may look weak and wretched from its infection, to nevertheless make itself smell alluring to pollinators.

> In making them more attractive to pollinators the virus gives these plants an advantage. ~ English plant pathologist John P. Carr

The virus doesn't hitch a ride on the bees that come calling for pollen, relying instead on its trusty aphids to porter it to the next plant. But the bees that spread the pollen around engender the next generation of plant to be more susceptible to infection. This virus plans ahead.

ꕤ Evolution ꕤ

> Nature is the source of all true knowledge. She has her own logic, her own laws; she has no effect without cause nor invention without necessity. ~ Italian polymath Leonardo da Vinci

2.5 billion years ago, after the continents had stabilized, prokaryotic sociality provoked a major evolutionary event. An archaean host and endosymbiotic bacterium committed to everlasting partnership through irreversible specialization. Thus arose single-celled *eukaryotes.*

The concept caught on. Within a few hundred million years, a variety of eukaryotic cell types had arisen, including plants, fungi, and the precursor to animals (metazoa). The biggest difference in lifestyle among them was diet.

> It seems that Nature has taken pleasure in varying the same mechanism in an infinity of different ways. ~ French philosopher Denis Diderot in 1753

With quantum exactitude, plants evolved the wondrous ability to turn sunlight and water into energy, and so were autotrophic. Though heterotrophic, fungi were flexible in eating ready-made foodstuffs from other organisms, dead or alive. In contrast, proto-metazoa were particular: requiring a diet of fresh prefabricated amino acids and vitamins.

Many unicellular organisms have a colonial form, often when food becomes scarce, or as a defense.

Single-celled green algae exposed to unicellular predators are easy prey. Clumping makes a difference. Small cell colonies offer an ideal tradeoff between security from predation and maintaining sufficient surface area for nutrient uptake.

The next evolutionary step was labor specialization at the cellular level. Prokaryotic colonial cells are not differentiated, but cells coordinate to perform different tasks.

Multicellularity was a déjà vu to the emergence of eukaryotes. Both were spurred by cellular cooperation.

Whence animals arose. In their innocence, early worms had no idea what trouble their inheritors a billion years hence would bring to their beloved home planet. In evolutionary terms, humans would turn *descent* into a dirty word.

The evolution of multicellular eukaryotes was by no means a declaration of independence. Quite the contrary. Eukaryotes have constant association with their prokaryotic forbearers in the form of a *microbiome.*

Animals are so dependent upon their microbial companions as to require their assistance in development, digestion of their foodstuffs, and protection from pathogens. An animal's microbiome is essential. In ignorant ingratitude, we call our most important friends *germs.*

> The human body and its symbionts can be viewed as a community of interacting cells. ~ American microbiologist and immunologist David A. Relman

In understanding evolution, a microscopic point of view makes a meaningful point. Genetic elements do not compete, nor do cells. Instead, the consistent themes to evolution, beginning at the cellular level, are: adaptation, cooperation, coordination, and specialization.

Multicellular life arose by *endosymbiosis*: the realization of mutual advantage by union. Photosynthetic chloroplasts and mitochondria, which are the power plants in plant and animal cells, were the products of this unification.

As life grew more complex, cells of all sorts were symbiotic add-ons. Greater complexity was achieved mostly through modularity. Once a structure is established, the potential exists for fractal repetition for different functionality.

> In modern complex organisms, novel adaptations result mostly from reorganization of existing structures. ~ Russian evolutionary biologist Alexander V. Badyaev

Organs are macroscopic extensions of the organelles within cells. The main innovation in plants was modularity: the evolution of semi-autonomous regions embodied within the embryonic meristems of roots and shoots.

¤✧¤

Evolution is the ongoing process of adapting to ever-changing environments. Those organisms that cannot do so quickly enough die out. That said, adaptation can be surprisingly swift.

♎ Overfishing ♎

> Human predators, by exploiting at high levels and targeting differently than natural predators, can generate rapid changes in both morphological and life history traits. ~ Canadian evolutionary ecologist Chris Darimont *et al*

In the last few decades of the 20th century, overfishing decimated fish stocks throughout the world. Plummeting populations from the pummeling also had dramatic evolutionary impact.

To gain an edge on being able to survive and reproduce, fish get smaller and grow up faster. This is a common strategy for animals experiencing rapid environmental change.

For example, Atlantic silversides can cut their average size in half in just 4 generations. Such brisk changes have been documented in populations of different fish across the world. Over the past few decades, the most commercially exploited fish have gotten 20% smaller, and their rates of living 25% faster.

The entertainment platform of existence is ever lively. Nature never sits still, nor does evolution, which generates astonishing adaptations that seem commonplace until you consider the specifics of what is going on. Then you see that exquisite design lays behind seemingly workaday traits.

♎ Frog Strike ♎

> For frogs, the forces acting on the tongue during impact and retraction can even be beyond the body weight of the animals. ~ German biomechanist Thomas Kleinteich & Ukrainian entomologist Stanislav N. Gorb

A frog uses its whip-like tongue to swiftly snag prey, faster than you can blink your eyes.* Its tongue can pull up to 1.4 times the frog's body weight.

A frog's tongue hits with a force five times that of gravity. Yet the food sticks as the frog swiftly snaps its bungee cord back.† The secret is the saliva, abetted by a clever tongue design.

Unlike other animals, a frog's salivary glands are not inside the mouth, where saliva drips onto the tongue. Instead, a frog tongue itself secretes saliva.

A frog's super-soft tongue curls around a prey upon contact, then deforms as it is pulled back towards the mouth. The tongue continuously stores the intense applied physical forces in its stretchy tissue, dissipating the shock via internal damping.

> The frog tongue acts more like a car's shock absorber than a pressure-sensitive adhesive; its viscoelastic nature enables rapidly applied forces to be dissipated in the tongue tissue. ~ American mechanical engineer Alexis C. Noel *et al*

The tongue is just one aspect of a frog's food delivery system. Special saliva is the other.

---

* A frog tongue strike takes only 20 milliseconds. Human eyes blink in one-tenth of a second; five times slower.

† Unlike the human tongue, which is attached at the back of the mouth, frog tongues are attached at the front. This lets them easily slip caught prey down their throats.

Frog spit has three distinct phases, superbly designed for the instant task at hand. The saliva is watery when it first hits its insect prey, filling all the bug's crevices.

In preparing for retraction, the saliva suddenly becomes sticky and thicker than honey, gripping the insect for the rocket ride back. Once the tongue is retracted, the saliva thins again, allowing the bug to be sheared off in the mouth.

In all phases, saliva thickness remains constant. Only its consistency varies.

A shear-thinning liquid, frog spit makes its wondrous conversions via quantum effects. What triggers the viscosity transforms is not known, but they appear as a byproduct of the tongue's extension.

Organisms evolve for an array of reasons: to better exploit resources, to improve breeding odds, to optimize efficiency, or to overcome an onslaught from predation or poison. The desire to live drives life and evolution. While adaptations may at times involve many changes, sometimes they are quite specific.

♎ Killifish ♎

Killifish are a hardy freshwater fish that often live in ephemeral waters: estuaries, wetlands, and vernal ponds that may mostly evaporate for a spell. Killifish eggs may survive weeks without water.

There are over 1,270 species of this once-abundant fish, found throughout in the Americas, and to a lesser extent in much of the world. But the pollution that humans produce can do in even the stoutest swimmers. Many millions of Atlantic killifish on the east coast of the United States lost their lives to a continuing chemical onslaught by human industry.

Yet some of these slippery slivers of silver survived the pollutants that plague American waterways. Genetic analysis revealed that several populations managed mutations which allowed them to withstand 8,000 times the levels of toxicity that might murder a lesser fish.

Though the genetic changes among surviving killifish were generally convergent, each population rapidly adapted to mounting toxicity in their own way. There was no other significant alteration in these killifish; only the ability to live in what would otherwise be toxic waters.

> Atlantic killifish populations have rapidly adapted to normally lethal levels of pollution in urban estuaries. Distinct molecular variants contribute to adaptive pathway modification among tolerant populations. ~ American evolutionary geneticist Noah M. Reid *et al*

There are countless examples of adaptation under duress. The only reasonable conclusion that might be drawn from these is that evolution can be purposeful (teleological); a concept cogently encapsulated in the word *adaptation.*

¤✧¤

Of course, other factors may affect the fortunes of populations, such as mating selection, which might alter the proportions of certain traits. But trait dominance does not lead to exclusivity. Instead, diversity typically persists. This is a strategy of Nature that provides a survival edge when changed conditions render a dominant trait disadvantageous.

♎ Guppy Love ♎

Male guppies display dazzling variations in their colorations. Female guppies have strong mating preferences based upon male color pattern.

This seems a formula for particular patterns to dominate the guppy world. Yet male guppies with rare color patterns persist, and do even better than males with more popular patterns; presumably because they are less likely to preyed upon by predators, who prefer to target the familiar.

Rare males have more matings and leave more offspring than those supposedly favored by females. This *rare-male effect* is the process in which the evolutionary fitness of a trait rises as its relative abundance decreases.

❀❀❀

Whatever competition occurs behaviorally, it does not diminish genetic diversity. While certain traits may dominate, mating selection does not winnow the gene pool, nor eliminate trait rarity.

Staying Fit

For our muscles to stay in shape we must exercise them. Rodents face a similar situation.

Even wild mice readily run on an exercise wheel. As with humans, rodent workouts are rewarded by a desirable dopamine deposit into their system, what people commonly call a "runner's high." Hence, though exercising is taxing, Nature provides an intrinsic incentive.

Migratory geese may fly thousands of kilometers at a stretch. To get themselves fit, geese simply sit on the water and stuff themselves with food. Yet by doing so, they develop stronger hearts and bigger flight muscles.

In a wide variety of organisms, seasonal signals prompt various alterations, including those related to fitness. Migratory birds undergo innumerable genetic changes that are stimulated solely by the changing hours of daylight.

Bear muscles do not waste away despite months of inactivity. Before hibernating, bears' bodies spontaneously release muscle-protecting compounds into their blood.

In needing to exercise to keep fit, rodents and humans are exceptions in the animal kingdom. Evolutionary tradeoffs prompted this. Our ancestors lived unpredictable lives, particularly regarding food supplies. Hence our body's ability to add fat more readily than any other animal.

Muscle mass is energetically expensive. Each kilogram of muscle takes 10–15 kilocalories a day to our resting metabolism. 40% of average human body mass is muscle.

> Most of us are spending 20% of our basic energy budget taking care of muscle mass. ~ American paleoanthropologist Daniel Lieberman

Our physiology evolved to let weight and fitness fluctuate depending upon food availability. Rodents face similar environmental conditions, and so have a selfsame set of traits related to musculature.

So, we see that muscularity is ultimately a matter of energy, not physicality. That muscles may tone themselves is evidence that our bodies are sustained by an energetic force, species-specific in its dynamic characteristics.

Saltation

> Natural selection acts only by taking advantage of slight successive variations; she can never take a great and sudden leap, but must advance by short and sure, though slow steps. If it could be demonstrated not by numerous, successive, slight modifications, my theory would absolutely break down. ~ Charles Darwin

A sudden leap in specific evolutionary development is termed *saltation*. Saltation dispels Darwin's gradualist "natural selection" hypothesis of evolution. Most saliently, saltation shows the creative flair of Nature in conjuring diversity. Many flowering plants, including the diverse variety of orchids, are the product of saltation.

♎ Saltational Scents ♎

Pheromones are particular blends of chemical compounds that are typically species-specific. Many organisms employ pheromones as signals to attract mates or other conspecifics, or for defensive or nefarious intent.

Altogether, there is an extraordinary diversity of blended scents, but with remarkable convergence of particular blends across very different life forms. Certain combinations are especially effective, and organisms zero in on them, irrespective of evolutionary descent.

> The high species specificity of pheromones suggests that there should be strong selection against small modifications in

these signals, and thus gradual evolution of pheromones through small changes in chemical components is unlikely. Instead, it seems more likely that pheromone evolution occurs via sudden major shifts in pheromone constituents. ~ Australian biologists Matthew R.E. Symonds and Mark A. Elgar

In examining the pheromones of bark beetles and fruit flies, Symonds and Elgar found "closely related species are just as different as more distantly related species. This ardently argues against the idea of minor shifts in pheromone evolution."

In drawing their inescapable conclusion of saltation, Symonds and Elgar wondered "how this mechanism of evolution actually works."

For the littlest ones, evolution is a matter of self-selection. Microbes carve their own evolutionary path by managing their own genetics.

Viruses, archaea and bacteria frequently decide how to adapt themselves to environmental conditions via self-generated genetic modifications. One way they do so is by sharing genetic concepts amongst themselves, a process called *horizontal gene transfer*.

These readymade adaptations can facilitate the rapid evolution of microbial populations. Thus, resistance quickly arises in populations of pathogens subjected to antibiotics.

At the organism level, adaptation visibly appears via bodily change. But physical changes are often the tip of the iceberg to functional transformations, including behaviors.

Predation has long been recognized as a key ecological factor for adaptive responses. But predation risk also drives the evolution of social complexity. Under the threat of predation, groups become more cohesive. Organisms stay together to minimize risk.

Sociality is not without costs, but these are far outweighed by the benefits. Thus, sociability is the norm for organisms throughout the tree of life.

Coral reefs are under severe stress from ocean warming and acidification. Only to a limited degree can coral quickly adapt on their own, via epigenetic changes. To better acclimatize, coral hire algal symbionts that can stand the heat. The coral gets improved temperature tolerance, and the algae find themselves safely harbored in a well-furnished home.

Evolutionary fitness is strengthened by cooperation. Life is not a hierarchy. It is an entangled web.

ঌ Whence Adaptation ଓ

The world doesn't change in front of your eyes, it changes behind your back. ~ English author Terry Hayes

Evolutionary biologists aim to answer a single question: how has the great diversity of life been productively generated?

The apparent increase in adaptive complexity of organisms over the history of life is one of the great mysteries of biology. ~ American evolutionary biologist Daniel W. McShea

While there are many common fallacies, there is no mystery. One common fiction is that randomness plays a part in evolution. Another is that evolution proceeds *without* objective.

Nothing in Nature is random. A provocation may be obscure, but it is there. Genetic mutations may occur from environmental pollution, but such genic contamination has nothing to do with evolution. Saltation and adaptation dispel the prospect that randomness is involved in evolution.

Many evolutionary biologists favor the idea of randomness because the implications of the alternative frighten them into befuddlement. Acts of Nature as intentional is termed *teleology*. Evolutionary biologists cringe from this concept because it puts natural causes outside the observable realm, and thereby makes the beating heart of Nature unscientific in their empirically-minded minds.

Biological evolution obviously exhibits intent: survival and reproducibility. The entire history of life is a dramatic presentation of organisms' will to survive and propagate.

That calamities may render adaptation awry only makes the show a spectacular melodrama.

No evolutionary biologist denies survival and perpetuation as driving forces. The objection is only the obvious ramification, thereby creating a paradox of denial in refusing to ascribe any impetus behind evolution.

⌑✧⌑

The blueprint manufacture of organisms is physically accomplished via *genetics*. From an evolutionary perspective, this field of study has been ripe with confusion between mechanics and process.

> Evolution is merely a reflection of changed sequences of bases in nucleic acid molecules. ~ English evolutionary biologist John Maynard Smith

Examining the microscopic mechanisms of DNA and its companions is nothing more than looking under the hood of Nature's vehicle to inspect the pistons. It does not explain the engine.

Functionally, *genes* are encoded molecules of designs for development. There simply cannot be a design without conceptual and compositional processes. This is the irrefutable teleology of Nature.

Convergent Evolution

Evolution is not mystical. Instead, its tendency towards particular traits makes its outcomes frequently foreseeable.

> Is evolution predictable? To a surprising extent the answer is yes. ~ Canadian evolutionary biologist Peter Andolfatto

The predictability of evolution is amply illustrated by *convergent evolution* (aka *parallel evolution*), whereupon vastly different organisms come to the same solution. Flight, vision, singing, parasitism, toxins, and camouflage are but a few of countless examples of convergent evolution.

Plants present a litany of parallel evolution. To begin, there were at least 5 independent evolutions of single-celled photosynthetic organisms into multicellular plant forms by

600 million years ago (MYA). Green algae alone gave birth to the land-based life commonly called *flora*, which became a prodigious potpourri.

Despite striking differences in climate, soils, and evolutionary lineages, plants share common needs, and means of achieving them. If one were to design an organism to optimize light collection, a branched structure with flat-bladed leaves might seem obvious. Nature created a wild variety that all aim at the same end, constrained in assortment only by biomechanics. A vast diversity of appearance belies functional convergence.

Kelp, a large seaweed that forms marine forests, has cells like phloem, the conductive tissue found in green land plants which transports sugars from sunlit tissue to those that live in perpetual shade.

As all plants are photosynthesizers, they have the same need to capture light, and internally exchange water and nutrients. Relationships in body plan geometry are crucial, and thereby constrained in terms of practical possibilities.

Over 350 MYA, competition for access to sunlight drove plants in very different families to reach for the sky. Vast forests of towering trees emerged during the Carboniferous period (359 – 299 MYA). Independent innovations for height continued since then. Many tall plants employ wood, but others, such as palms and bamboo (an ambitious grass), do not.

Acquiring carbon dioxide for photosynthesis through controllable pores (stomata) causes water loss by transpiration. Although plants evolved various mechanisms to alleviate this problem, one well-established solution is C_4 – a technique that independently evolved at least 45 different times, starting 32 MYA, and most recently 4 MYA.

ꟿ Plants ꟿ

Plants appeared on land over 500 MYA. The way had been paved by hardier life.

The first soils were prepared by microbes, algae, and lichen that had arrived much earlier. They broke down rock to sustain themselves, releasing minerals valuable to vegetative growth.

Plants played out what their predecessors had started. Rock-hugging mosses extracted vital minerals from the substrate upon which they were perched, causing chemical weathering on the Earth's surface. These terrestrial pioneers paved the way for a richer life for their descendants.

The earliest plants had help from microbes that garner minerals from the soil. This grew into mutual relations. Plants today cultivate specific root microbes when growing in nutrient-poor soil.

♎ Ferns ♎

From humble beginnings, plants evolved with verdant flourish. Ferns emerged over 360 MYA.

Ferns were initially quite successful. Like sharks, which had a divine design that kept them in good stead for hundreds of millions of years. The evolutionary advance of ferns was modest for 180 million years.

Then calamity struck. Unlike sharks, ferns could not compete with more modern designs. The rise of towering trees and flowering plants spelt their demise.

Desperate for an innovation to save them from the darkness of extinction, ferns found the answer in learning to live in the shadows of more advanced plants. Moving forward took looking back.

Ferns picked up a gene from an earlier-evolved plant – hornworts – that let them thrive on shady forest floors. From 180 MYA, the one lineage of ferns that had managed to survive proliferated into 12,000 species.

Plants are master molecular constructionists. They can concoct formulas for every purpose: sugary confections to entice cooperation, scents to deceive, and poisons to ward off predation.

♎ Goldenrods & Gallflies ♎

Goldenrods are especially bothered by gallflies, whose entire life cycle is centered around the flowering plant. But the goldenrod has figured out how to put the parasite off.

A goldenrod sniffs out when a male fly is about. Yes, plants can smell.

That little parasite is probably trying to find a mate: sitting atop a leaf or bud, dancing when a female comes into view. Such insolence is not to be borne. The goldenrod rolls up the welcome mat by producing toxins that deter egg laying.

Goldenrods offer one of innumerable examples of inscrutable knowledge behind adaptation. Only by Nature's grace could a plant possibly know what molecular combination effectively deters egg-laying by its nemesis. Many plants produce compounds (secondary metabolites) that specifically target molecular mechanisms essential to a tormenter's development, metabolism, or reproduction.

> Of course it makes sense that evolution has used all the available opportunities to enhance plant fitness. ~ English botanist Beverley J. Glover

In studying the history and web of life, an intelligent force behind adaptation is everywhere apparent. The combinations by which life's goals are met are often subtle and complex. Flowers' mastery over physical forces illustrate.

Flower Physics

> The Earth laughs in flowers. ~ American poet Ralph Waldo Emerson

Attracting pollinators is essential to the reproductive success of flowering plants, which evolved with their animal admirers in mind. A flower's corolla (petal structure) is a visually-arresting billboard that titillates with its promise of nectar and pollen rewards within. A pleasing scent adds to the allure. The petals themselves offer easy-grip textures or other inducements to visitors.

Chemistry has a lot to do with these floral multimedia presentations. But plants also masterfully use subtle tricks of physics to achieve irresistibility for their flowers, and tilt the transactional bargain in their favor.

Light Show

The color of most objects comes from chemical pigments that selectively absorb certain wavelengths of light. Flowers use pigments to provide contrast with the surrounding green foliage.

But there are also physical means to generate color, or, at least, to modify the color that would be produced by a pigment alone. Plants generate dazzling optical effects by exploiting light's interaction with microscopic structures in their flowers.

The snapdragon is a flowering plant native to the Mediterranean. The name comes from the flowers' reaction to having its throat squeezed: the front of the flower snaps open like a dragon's mouth.

The snapdragon differs from its close relations by having a genetic modification that makes the cells on the surface of its flowers conical rather than flat. This conical geometry

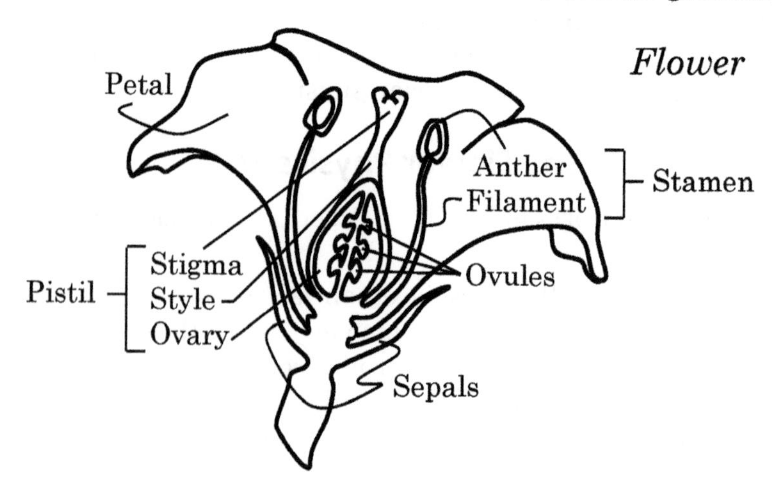

acts as a lens: focusing light into the vacuole containing the flower's pigment, while also scattering more light reflected from the mesophyll (the interior of the leaf).

The effect is a more intense coloration on a petal that sparkles. By playing with the light, plants are better able to strut their wares to potential customers at a distance.

Snapdragons with conical flower cells receive more pollinator visits, and produce more fruits, than those with flat cells. As to other species, conical cells are present on most flowers, especially on the parts directly exposed to pollinators.

Having a smooth, lustrous surface is a good way of being conspicuous. But this can be improved upon by careful accenting. Some flowers add touches of gloss on petal tips that glitter when light hits at a certain angle.

Iridescence is another optical trick that flowers employ. Lustrous changing colors are achieved by microscopic structures on or just below a petal's surface which generate color by diffraction and/or interference.

Flowers achieve iridescence via striations that create disordered diffraction gratings. Floral iridescence acts as a cue to pollinating insects. Disorder can have adaptive value.

Insects use their sensitivity to polarized light in several ways, from general navigation to finding food and nesting sites. Knowing this, flowers often display polarization patterns that emerge from micro-differences in geometric surface structure.

Mechanical Tricks

It's not easy to land on a flower moving in the wind, and then find the correct approach angle to access the reward. With this in mind, many plants evolved strategies that take advantage of the interplay between pollinator, flower surface, and gravity, to limit floral access to certain groups of animals.

A simple way in which many flowers improve the grip and handling efficiency of their flowers is to use conical cells on the petal. These improve foraging efficiency for bees by providing an interlocking surface for their tarsal claws, which is particularly important when flowers are handled at difficult angles, or in windy or slippery conditions.

Some flowers manipulate pollinators by selective loss of these cells. Several plants on Macaronesia independently dropped use of conical petal cells when transitioning to avian pollination.

This evolutionary change makes flower petals more slippery to insects, thus minimizing their nectar robbing. Even bee-pollinated flowers selectively distribute their epidermal cell types to control the physical forces acting on foragers.

Plants use physical forces to aid pollen transfer, particularly placing pollen on specific parts of a pollinator's body. Controlling pollen placement is critical for plants that flower in the same habitat as other species that use the same pollinator, as accurate placement limits the chance of interspecific pollen transfer, which is wasteful.

Impatiens frithii, endemic to Cameroon, is a small, inconspicuous plant when not displaying its bright red flowers. *Impatiens* manages specific pollen placement for its pollinating sunbird by a twist in its flower's nectar spur, which curves upward. When the bird inserts its bill to feed, it exerts physical force that causes the flower to rotate 180 degrees, placing pollen below the beak. To achieve optimal pollen placement, the flower has a flexible stalk that can tolerate the anticipated rotation by the bird.

Trigger plants play an even more impressive mechanical trick. The male and female reproductive organs are fused into a floral column which acts as a sensitive trigger: snapping down within milliseconds when a pollinator alights, covering the insect with pollen.

The column recovers its original position over the following few minutes. Movement is driven by changing cell volume and length via potassium ion transport. To prevent self-pollination and ensure cross-pollination, the male (anther) and female (stigma) organs take turns dominating floral column function.

Slower movements can also prove highly effective at pollen placement. Stamens of the prickly pear cactus move inwards over the course of 2 to 20 seconds after being touched by an insect of appropriate weight. The movement forces a pollinator to push past the anther to exit the flower, increasing pollen distribution.

Electrostatic Forces

Plants also harness electrical charges to improve pollination. Pollinators accumulate an electrostatic charge from flight. This induces an electrical field with flowers, which are also charged.

The field strengthens as a pollinator approaches, facilitating pollen transfer from the anther to the insect or bird, and from the pollinator to the stigma. Foraging bees sense this field, and use it to determine whether a flower is worth the trip. A distinctive field lingers on a flower after it has been visited, thus indicating its reward status.

Plant Deception

> Honesty may be the best policy, but it's important to remember that apparently, by elimination, dishonesty is the second-best policy. ~ American comedian George Carlin

The botanical kingdom is rife with beguilement. 5% of flowering plants entice pollinators via various ruses. Others deceptively attract insects for their own consumption.

Carrion flowers smell of rotten meat to attract scavenging beetles and flies, which it then slathers with pollen. Passion vines, beloved by some butterflies as food for their caterpillars, have yellow spots on their leaves that look as if eggs had been laid by a pregnant female. Numerous carnivorous plants lure insects with sweet odors only to devour them.

Deception plays an important role in plant defense. Plants having their leaves chewed on by insect pests emit a chemical cue which tells the insect that the plant is damaged and a poor source of food. These airborne missives are noticed by neighboring plants, warning them to prepare their own chemical defenses. To be forewarned is to be forearmed.

In dense populations, deceptive chemical defenses keep insect herbivores on the move. Distributing the damage helps a community survive.

Plants can tolerate modest injury without affecting their fitness. Goldenrods, for example, can tolerate losing up to 30% of their leaves. Anymore and they are goners. The will to live is lost.

♎ Parachute Plants ♎

The parachute plant has cone-shaped flowers which trap insects that come to its blooming parlor, letting them escape only after the flower has wilted. They leave covered in pollen.

Freeloader flies are particularly attracted to parachute plants. These flies have a predilection for lapping up the vital fluids that leak out of honeybees after being bit by spiders. Parachute plants lure freeloaders by concocting the scent of a dying bee. If you wonder how a plant knows what a dying bee smells like, you're on the right track.

Many plants have subtle mechanisms that promote production of protective compounds only when needed. They know what time it is, and the time of day that common pests rouse themselves to bring ruin. In anticipation, plants ramp their defenses in likely locations of assault.

¤✧¤

Plants provide the pharmacy for animal life. Many animals self-medicate with herbal remedies.

Woolly bear caterpillars carefully select and consume pomegranate leaves rich in an alkaloid that kills parasitic fly larvae laid inside their abdomen.

Fruit flies face a similar problem. Parasitic wasps are prone to lay eggs into fruit fly larvae, a fatal fate for a fledgling fruit fly. But fruit fly larvae know when they are infected. The larvae seek out high alcohol-content fruit, imbibing the boozy fruit to kill the parasites.

Fruit fly moms look out for their little ones by preemptively laying eggs into alcohol-laden food if a parasitic wasp is spotted in the neighborhood. This precaution protects larvae when they hatch.

Primates know numerous medicinal remedies. Capuchin monkeys eat pepper plant leaves as an antiseptic. They also know that nutmeg fruit has antimicrobial properties.

Chimpanzees treat intestinal worms by ingesting the leaves of plants with anti-parasitic compounds, such as wild sunflower.

Plants have ever been the foundation of human medicine. Before laboratory synthesis developed in the late 1800s, 80% of the substances used to cure diseases were plant-derived. Plants still account for some 40% of the drugs taken.

Directly or indirectly, all animals depend upon plants for their survival. Plants are the builders of ecosystems which most land animals inhabit. The more abundant and diverse plant life is in an ecosystem, the more vibrant the living environment is for all that live there. Conversely, a paucity of plant life offers only the meanest existence to animals.

Intelligence

> Life requires cognition at all levels. ~ American molecular biologist James A. Shapiro

Every life is a consciousness housed in a mind-body. Though organisms may not have a physically identifiable seat of intelligence, such as a brain, all have a mind. This is shown by the way organisms behaviorally adapt to their habitat.

> Cognitive abilities are found very low on the evolutionary tree. ~ Austrian neurobiologist Gero Miesenböck

¤✧¤

Just as energy fabricates matter, the mind depicts the body. Feeling a phantom limb illustrates. 60–80% of amputees feel sensations where amputated limbs used to be. Animals other than humans also sense phantom limbs.

Mind-body asymmetry is also illustrated by stress. Physical exertion tires the body, but does not otherwise tax awareness, which may be enlivened by exercise. But mental stress, such as worry and fear, systemically degrades both mind and body.

♎ Quorum Sensing ♎

> Bacteria show patterns of collective behavior that reflect social intelligence. ~ Israeli physicist Eshel Ben-Jacob *et al*

Using language compellingly is commonly thought to be unique to humans. But persuasive conversation started billions of years ago with bacteria.

Quorum sensing (QS) is the general term for decision making in decentralized groups to coordinate behavior. Microbes practice quorum sensing by exchange of chemical signals. Using a common language, many different bacteria employ quorum sensing to synchronize their activities. Viruses employ QS to make group decisions during infection.

Such language is widespread. Eukaryotic cells respond to QS signaling. Human white blood cells can be induced to change their behavior by receiving such signals.

> Quorum sensing is used to coordinate the switching on of social behaviors at high densities when such behaviors are more efficient and will provide the greatest benefit. ~ English molecular biologist Sophie E. Darch

For microbes, the density of group populations must be high enough for QS to be effective in coordinating activities. Until population density reaches a recognized threshold, QS is merely a monitoring mechanism. Biofilms (colonies of microbes, commonly called *slime*) facilitate productive quorum sensing.

Many species of bacteria coordinate their gene expression via QS. In effect, via quorum sensing, single-cell microbes behave as a multicellular organism.

Quorum sensing is a conserved trait throughout life. Social insects use quorum sensing to make collective decisions, such as where to forage or nest, as do schools of fish when feeding or evading predators.

∞ Microbes ∞

We know so little of the little ones. Viruses, archaea, and bacteria are far too successful to mark it down to dumb luck.

Microbes navigate and adapt to their surroundings. They communicate at the molecular and genetic levels, amongst themselves and with other species.

> Microbes indulge in a variety of social behaviors involving complex systems of cooperation, communication, and synchronization. ~ English microbiologist Stuart A. West

Microbes acclimate themselves to other species, and accommodate, or antagonize, with various responses. Through selective genetic exchange, microbes demonstrate what approximates to empathy, and, when times are tough, work for the betterment of their community.

♎ Altruistic Algae ♎

When presented with an abundance of nutrients, single-celled algae bloom into a prolific population. As the food runs out, individuals self-selectively commit suicide to sustain others. The nutrients from self-sacrificing algae can only be used by relatives, and inhibit the growth of non-relatives. Not only does suicide help kin, it can also harm competitors.

∞ Plants ∞

> Plants are very sophisticated chemical factories; able to produce thousands of different compounds, each one presenting unique biological properties. ~ Swedish botanist Stefano Papazian

Plants live a life of conscious chemistry. Their thoughts and behaviors are exercises of molecular awareness. The contrast to animals is incomparable.

Deciding priorities and energy allocations is so enormously complex that no plant behavior is autonomic. There is no plant unconsciousness.

One aspect of existence that is the same for both plants and animals is memory. Plants remember their ecological interactions and derive meaning from them.

♎ Choosy Consumers ♎

Sundews and the Venus flytrap are carnivorous plants that exercise discretion in food selection.

Sundews have leaves covered in long glandular hairs that secrete a sticky mucilage. Any insect landing on them becomes enveloped in the adhesive gum. All the leaf tentacles gradually bend over to ensure an entrapped prey, whereupon digestive enzymes are released and dinner is served.

The Venus flytrap has V-shaped leaves which shutter any creature careless enough to forage for food there. It too lets loose juices to chemically cook and consume its hapless visitors.

Perchance a bit of debris strays on a sundew or flutter a Venus flytrap shut, the plants sense they have caught something undesirably indigestible. A flytrap quickly reopens for business. A sundew loosens its grip to entertain something savorier. Neither bother to wastefully release enzymes when there is nothing good to eat.

Sundews are a less patient sort than Venus flytraps, which are content to await their victims. If a sundew leaf smells a meal within reach, the leaf bends towards it until it is within a tentacle's grasp.

❁❁❁

Plants have evolved complex sensory and regulatory systems that allow them to modulate their growth in response to ever-changing conditions. ~ American botanist Daniel Chamovitz

Plants sense their environment, and respond by adjusting their activity, morphology, physiology, and phenotype. They account for their resources and plan accordingly.

Lianas are climbing vines that root in soil and send themselves skywards. They will not attach themselves to particular trees, even if the opportunity presents itself. The trees lianas refuse are those least suited to their climbing style and objective: smooth trunks with umbrella tops that won't do them much good when the reach the canopy.

¤✧¤

The basic tradeoff for a plant is between growth and defense. Spending energy on defensive measures limits growth potential. But defensive strength cannot be suddenly amassed. So, the tradeoff between growth and defense requires an energy budget, which is meticulously decided as an ongoing process.

Growth itself is a decision-laden process. A plant must decide how best to allocate its resources amidst an incredible diversity of options, such as root, stem, leaf or bark growth.

> Because plants cannot run away from danger, they have evolved defenses against pathogens and herbivores that rival and even exceed the sophistication of many animal immune systems. ~ American behavioral ecologist Andrew G. Zink & Chinese botanist Zheng-Hui He

Plants possess a wide repertoire of defenses and healing remedies. By recognizing signature molecules of microbial malevolence, plants are actively aware of an infection, and consciously decide how to deal with it. Among other options, they might decide to sacrifice a region around the infected area, to prevent its spread.

Plants must constantly assess the probabilities of favorable outcomes given an ample array of possibilities, a cognitive skill known as *risk sensitivity*. Experience and calculation of relative gain determine decisions in plants just as they do in humans and other animals.

In one set of experiments, pea plants were given the opportunity to select, via root growth, areas which either had a consistent amount of nutrients or fluctuating food levels. When nutrients were abundant where they were, plants played it safe, opting for consistency. But when living in suboptimal soil, plants preferred to take a risk.

Those flowering plants with both female and male sex organs recognize and reject their own pollen, thereby averting inbreeding. With intricate genetics behind it, considerable cognition is involved in this process.

Plants solve problems and meaningfully communicate with their neighbors and other species in the appropriate molecular language. Plants understand and manipulate other species chemically in a variety of ways, from soliciting and assisting desirous cooperators to thwarting and killing those with nefarious intent. Human intelligence is puny compared to that of flowering plants, which have an earthy savvy that has been honed over the past 245 million years.

꧁ Animals ꧂

> The more we look at the behavior of insects, birds, and mammals, including man, the more we see a continuum of complexity rather than any dramatic difference in kind.
> ~ American ethologists Carol G. Gould & James L. Gould

Animals originated during the chilly Cryogenian period. The last common ancestor of animals arose nearly 800 million years ago.

The term *animal* comes from the Latin *animalis*, meaning "having breath." But breath does not distinguish animals from plants. Plant pores have a regulated cycle of breathing in carbon dioxide and exhaling water vapor.

Unlike other life forms, animals evolved centralized intelligence processing centers for digestion and cognition. In later-evolved animals, identifiable brains exhibit electrochemical activity that correlate with mental processing.

Such entanglement between energetic fields and seemingly physical particles is how the weave of illusory duality is attained. It should be unsurprising that such intricacy transpires in the processing that generates phenomenality.

Brains & Neurons

> Brains and neurons obviously have everything to do with consciousness, but how such mere objects can give rise to the eerily different phenomenon of subjective experience seems utterly incomprehensible. ~ American biologist H. Allen Orr

The religious error of positing the physical as the penultimate of existence is compounded by neurobiologists, who wrongly identify neurons, not glia, as the cells of cognition.

> As long as our brain is a mystery, the universe, the reflection of the structure of the brain, will also be a mystery. ~ Santiago Ramón y Cajal

In the late 19th century, Spanish neurologist Santiago Ramón y Cajal elucidated and fiercely defended what became known as the *neuron doctrine*: neurons were the cells of intelligence. Cajal shared the 1906 Nobel prize in physiology and medicine with Italian physician and pathologist Camillo Golgi, who had identified astrocytes, a glial cell type, as important in thought processing.

Despite Golgi's findings, Cajal's thoroughly neuron-centric influence prevailed, becoming the mainstream school of modern neuroscience. Because of him and his followers, glia went unstudied for six decades.

> Until recently, our understanding of the brain was based on a century-old idea: the neuron doctrine. This theory holds that all information in the nervous system is transmitted by electrical impulses over networks of neurons linked through synaptic connections. But this bedrock theorem is deeply flawed. ~ American neuroscientist R. Douglas Fields

Misattribution is made by facilely mistaking coincidence for cause. One cannot see what is not looked at, or looked at with a predetermined perspective. In academic disciplines, only economics has proceeded with as much built-in bias as neuroscience.

Neurobiologists long assumed that neurons were the governors of consciousness, particularly the transition between sleep and the awake state. Instead, that physiological transition occurs through ion flows regulated by glia.

Hunger is monitored and its response controlled by glia, not neurons. This is done by regulating the release of the hormones which control the sensation of hunger and adjust energy expenditure. The effects of metabolic hormones come by commanding neural circuits.

In all degenerative brain diseases, the first symptom, even before the loss of mental faculty, is losing the sense of smell. Smell receptor cells actively lock onto ambient molecules for detection. This requires frequent replacement of these receptors. Hence, sense of smell is constantly changing, and an apt indicator of holistic health.

The olfactory bulb, locally responsible for smell, has the highest turnover of cells in the brain. Glia are the stem cells for this turnover.

Neural processing cannot explain the pattern matching that predominates mental processing. After extensive study of nerve cells for well over a century, neurobiologists still cannot explain memory via neurons. That's because mentation physiologically transpires in glia, not nerves. Cognitive diseases, such as epilepsy and autism, result from defective glia, not neurons. In old age, Alzheimer's disease comes via crippled glia, not worn-out neurons.

A long-known fact is that brain tumors are almost always glial cells. These tumors would not be so devastating if neurons were running the show.

Glia are the adult stem cells in the brain. They reproduce themselves, and produce neurons if need be.

Glia manage nerve cells. Glia guide developing neurons, sop up chemicals used in cell-to-cell communication, and generally contribute to the health and well-being of nerve cells and their environment. But glia do much more.

Glia regenerate and grow locally in order to store more information. Via intercellular calcium waves, glia distribute and process information. Neurons have no memory capacity beyond those of other cell types. That's because mentation has its physical correlate in glia, not nerves.

The explosive growth of the human brain in the first year after birth owes to astrocyte propagation. Meantime, nerve cell growth is fractional.

Children begin to experience dreams and are able to retain long-term memories by around age four, after glia grow and establish themselves postnatally. If neurons held memories, people could recall being in the womb.

Nerve cells predominate in the cortex, which is gray matter. Cortex development increases to about age 8, then the brain becomes more streamlined. An adult cortex is considerably smaller than that of an 8-year old.

Learning results in a temporary increase in neurons in the affected area. As the learning takes hold and becomes rote, neural pathways streamline, with neurons atrophying and lessening in number. But more glia grow and remain robust with learning.

The cortex thinning that occurs from childhood is mostly neuron loss. Autism arises with a failure to prune neurons. In contrast, smarter children experience accelerated neural thinning.

> More is not better when it comes to synapses, for sure, and pruning is absolutely essential. ~ American molecular biologist Lisa Boulanger

An evolutionary perspective highlights the importance of glia. If glia function as the mental library, then species with greater cognitive facility should have proportionally more glia. And so it goes.

The ratio of glia to neurons increases with what is broadly considered cognitive capacity. 3% of a leech's intelligence cells are glial. The rest are neurons. In an earthworm, glia make up 16% of the nervous system. That ratio rises to 20% in flies; 60% in rodents; 80% in apes; and 90% in humans.

> Glia are the conductors. ~ R. Douglas Fields

Actually, glia are *not* the conductors. They may rule the roost with regard to intelligence physiology, giving a deceptive appearance of conducting. But the material ruse frays when trying to explain how cogent mentation could possibly arise from disparate electro-chemical reactions.

This difficulty may be glossed over when massive tissue is involved; simply hand-wave that so much meat matters. But when brains become tiny and cognitive skills stay sharp, the jig is up.

> Experiments with insects and crabs have demonstrated their remarkable ability to learn and memorize complex visual features. ~ Argentinian neuroscientists Julieta Sztarker & Daniel Tomsic

♎ Seashore Crabs ♎

Though often crowned with expansive Latin titles, crabs are not noted scholars. *Chasmagnathus granulatus* is a teensy seashore crab. It leads an idyllic life: digging into the sand for food, and doing its best to avoid becoming a meal for a diving seagull, the crab's nemesis.

Such a seemingly simple lifestyle belies survival needs that conjure considerable cognition. Burrowing crabs have excellent long-term memory – both for good feeding spots, and places where seagull attacks are more probable. These crabs are also able to discriminate between real seagulls and look-alike decoys.

> These animals don't have millions of neurons like mammals do, but they can still perform really complex tasks. ~ Julieta Sztarker

This burrowing crab's brain is smaller than the point of a pencil; a bitty brain with a tiny fraction of physical substrate to process mentation compared to mammals or birds. Yet the crab's mental powers are undeniably sophisticated, quite up to the demands of its lifestyle and then some.

Researchers watched the electrical activity of brain neurons in *C. granulatus*, and came to a fantastic conclusion.

> The behavior of the crab was found to be accounted for by the activity of a small number of neurons. ~ Julieta Sztarker & Daniel Tomsic

The scientists involved believed that "a small number of neurons" account for "really complex" behaviors. You would be sensible to be skeptical of such a sophistic denouement – a wise crab, wary of the seagulls of pseudo-science.

Precocious Knowledge

> Precocious knowledge can help naïve individuals in making correct predictions and deciding whether to approach or avoid and object, and how to cope with a situation encountered for the first time. Evidence of precocious knowledge has been documented in species with a short life span, where learning by trial and error could be too costly. ~ Italian evolutionary biologist Elisabetta Versace & Italian cognitive psychologist Giorgio Vallortigara

For animals, a primary evolutionary dichotomy exists in developmental strategy that determines innate knowledge. This trade-off (life-history variable) involves the degree of maturity that an animal has when it begins its life.

An animal may be precocial or altricial. Newborn precocial animals come into the world equipped to cope independently. By contrast, offspring of altricial species are dependent upon parental care to survive, as they are born or hatched in an immature form. Orangutans and humans are altricial to an extreme.

These two modes of development evolved based upon food availability and predation pressure. Birds exemplify.

Female precocial birds must obtain abundant resources before laying, to produce the energy-rich eggs needed to support greater in-egg development of chicks. Eggs of precocial birds may have twice the calories per unit weight as those of altricial birds.

Altricial avian females do not have such large nutritional needs before egg laying; but, with help from their mates, they must be able to find enough to feed their fledglings.

While in the nest, an entire brood is extremely vulnerable to predation, and so dependent upon concealment and parental defense. In contrast, precocial birds quickly leave the nest with some ability to avoid predators. There is much less chance of an entire brood of precocial chicks being killed.

In altricial species, embryonic development is relatively rapid. The neonatal brain will grow from its small size after birth. In contrast, development before birth is longer for precocial species, and the neonatal brain larger. But there is no consistent difference in adult brain size between altricial and precocial species.

The difference between precociality and altriciality is especially broad in birds. Precocial birds open their eyes upon hatching, are covered in down, and leave the nest within a couple of days. Chickens and several water birds, including many ducks and geese, are precocial.

Some birds are superprecocial. Megapodes are stocky, largish, chicken-like birds, endemic to Australasia. They hatch with a full set of feathers; some can fly on that same day.

Altricial chicks hatch with their eyes closed, covered in little or no down, are incapable of leaving the nest, and must be fed by their parents for an extended period.

All passerines, which comprise half of the bird world, are altricial. Passerines include most perching birds.

Parrots have the best of both worlds. They are altricial; but parrot eggs are nutrient-rich, like those of precocial birds.

Parrot life history is like humans. Both are highly intelligent, born with eyes open and large brains. But developmental success is predicated upon significant investment in parental care.

Like megapodes, some mammals are highly precocial. Common wildebeest calves can stand with minutes of being born, and walk about within 30 minutes. Within a day, a wildebeest can outrun a hyena.

Precociality gives this wildebeest (aka *blue gnu*) a great advantage over other herbivores. Blue gnus are 100 times more abundant in the Serengeti ecosystem where they live than their closest relative, hartebeests. Hartebeest calves can take up to a half-hour after birth before they can stand up, and are unable to keep up with their mothers until they are over a week old.

A key aspect of precociality is being born with the wiles to deal with a dangerous world. Precocial animals are certainly capable of learning, but they come equipped with considerable knowledge upon birth. How is this possible?

That genes seemingly encode various regimes for physical development is well known; though exactly how they do that remains enigmatic, as genes are merely templates for proteins and other organic macromolecules. We also know that behavioral traits are inherited (another genetic mechanics mystery). Considering the evolutionary requirements of precociality, there must be some way that information itself is imparted to newborns.

> Solid evidence shows the existence of unlearned knowledge in different domains in several species. ~ Elisabetta Versace & Giorgio Vallortigara

Though we know that savviness can be innate, there is nothing to support the idea that genetic matter manages to encapsulate expertise. Besides lacking evidence, it is simply inconceivable that nucleic acids store knowledge which consists of actionable concepts and categories pertinent to the outside world.

There is no way to explain the cognitive abilities of precocial animals by material means – within genes. Consider instead that a species-specific knowledge set is energetically embedded in an embryonic consciousness coupled to a mind-body, and precocial knowledge makes perfect sense.

Genetics put on quite a show at the molecular level, but like so much else, the story of existence is incomplete until the intangibles of Nature are factored in. An intelligent force of harmonious structuring will never appear under a microscope; but it most certainly is here, there, and everywhere.

༄ Humans ༄

A human neonate naturally assumes a oneness with the world. Duality gradually dawns.

Around 3 months, infants begin to distinguish between people, objects that may be biological, and moving inanimate objects. But babies do not recognize themselves in a mirror until they are into their second year of life.

Only at around 2 years do children become aware of a distinction between thoughts in the mind and things in the world. The concept of self develops around this time.

> Infants lack a concept of the self. The emergence of self-conscious emotions is a consequence of the concept of self. ~ American psychologist Lisa J. Cohen

2-year-olds also start to understand emotional contexts: that they, and others, are happy when they get what they want, and sad if not. They also begin to appreciate that there may be a difference between what they want and someone else wants.

The temper tantrums of young children occur not only because of frustration, but with outrage that frustration even exists.

> In these moments, children are enraged that they should have to be frustrated at all, that their will can actually be thwarted. ~ Lisa J. Cohen

A 3-year-old is apt to speak of what people think and know. These observations are of course limited to what the 3-year-old knows.

A crucial cognitive development occurs around the age of 4 years, when children realize that thoughts in their mind may be false.

For example, a child may discover candy in a familiar pencil box. After this discovery, ask a 3-year-old what their friend will think is in the box before looking, and she will think her friend knows what she knows: candy. At 4, the child will understand that a friend would be tricked, as she was.

Whereas 3-year-olds also do not remember that their belief changed, a 4-year-old recalls the assumptive self-deception.

By 4–5 years, children realize that people talk or act on the basis of how they *think* the world is, even though their thoughts may not reflect how the world *actually* is. With such awareness, 5-year-olds will not be surprised if a uniformed friend looks for pencils in the marked box which now has candy.

These are two crucial cognitive developments. First is the discovery of *abstraction*: that the mind creates its own world, distinct from reality. The second is *theory of mind*: that others have minds which are different than one's own.

In misconstruing others' thoughts and intentions, theory of mind is a projection fraught with the potential for self-deception. It is also the preeminent skill for sociality: to sense what someone values, and their worldview.

> The most important development in early childhood social cognition is the development of theory of mind. ~ Canadian developmental psychologist Janet Wilde Astington

Theory of mind is the basis for empathy. Coupled to theory of mind, the value of abstraction provides the basis for deception: that one can convey a convincing illusion.

Humans are not the only ones who deceive. Many plants and animals practice deception to defend themselves, or as a lifestyle. Femme fatale female fireflies deceptively attract males of another species for a meal, with the duped males as the main course.

The Unreasonable Power of the Mind

Philosophic materialists who blabber that the brain does our thinking for us have no explanation for how the mind manages its feats. Instead, a massive body of evidence indicates that the brain is just for show.

> The fundamental problem is that our brain doesn't work in real-time. The brain actually works rather slow. ~ American psychologist Gerrit Maus

♓ Immunity Images ♓

Mere visual perception of disease-connoting cues promotes a more aggressive immune response. ~ American psychologist Mark Schaller *et al*

Simply seeing sick people from a distance, or just looking at photos of those ailing, kicks the immune system into heightened alertness. How such imagery gets translated into mustering the body's cellular defense system cannot be explained physiologically.

It makes evolutionary sense that the immune system would respond aggressively only when it's needed. ~ Mark Schaller

This remarkable capacity we possess to understand something of the character of another person, to form a conception of him as a human being, with particularly characteristics forming a distinct individuality, is a precondition of social life. ~ American social psychologist Solomon E. Asch

Studies have repeatedly shown that people make accurate assessments of trustworthiness, competence, political affiliation, and other traits, within a fraction of a second. All that is needed is a glimpse of a face.

We can accurately judge a person's honesty in only a tenth of a second. ~ English psychologist Simon J. Makin

Though we take it for granted, there is no accounting for how our minds manufacture so much from so little. The mind makes bold generalizations, constructs elaborate abstractions, and builds robust causal models surprisingly quickly from inputs that are sparse, noisy, and ambiguous; in every measure, far too limited.

A massive mismatch looms between the information coming in through our senses and the outputs of cognition. ~ American cognitive scientist Joshua B. Tenenbaum, American psychologists Charles Kemp, Thomas L. Griffiths, and Noah D. Goodman

Generalization from sparse data is central to learning in several areas, notably language. Terminology, morphology, and syntax are learned from meager references.

2-year-olds can learn how to use a new word from just an example or two. Because they can use a new word appropriately in a novel situation, we know that young children grasp not just the sound, but the meaning and the context, thus generating a comprehensive conceptualization.

Viewed as a computation on input information, this is an astonishing feat. In understanding new words, how a child grasps the boundaries of objects or actions from so few examples is inexplicable.

The two basic lessons of statistics are that sample size dictates correlation quality, and that correlation does not imply causation. Yet children routinely, and correctly, infer causal links from a few events; far too few to even compute a reliable correlation.

The deepest intellectual accomplishments are the constructions of knowledge systems. Overarching theories of physics, psychology, and biology are inferred on scales far surpassing the facts on which they are based.

The mind's algorithms are clearly probabilistic. Generalizations from examples appear to be structured as representations in various mathematical forms: clusters, rings, trees, grids, and directed graphs, to name a few. Learning progresses from simpler structures to more complex.

Children initially assume exclusive clusters when learning words. Only later do they discover that nomenclature has a tree-like hierarchy.

Science advances similarly. Biologists have long sought to systematically categorize the multitudinous forms of life on Earth. In 1866, German biologist Ernst Haeckel, inspired by Darwin's work, laid the foundation for modern biological taxonomy by proposing a tree of life rather than the existing linearity. The tree concept was taken up by Swedish biologist Carl Linnaeus, who fumbled the specifics rather spectacularly with guesswork.

Biological classification progressed from how organisms looked to how they may have speciated through time. This involved distinguishing *clades*: groups based upon evolutionary descent. The ability to analyze genetics greatly helped the effort.

This cladism approach revealed that life's diversity was much more sophisticated than could hang on any tree. But no more sophisticated structure could be arrived at. Despite astonishing ability to jump to correct conclusions (at least sometimes), our conceptual acumen is limited.

In 1869, having noticed patterns in the properties of chemical elements, Russian chemist Dmitry Mendeleyev published the modern form of the periodic table. While illuminating, this grid only begins to capture the wizardry by which molecules exist; a complexity which has defied embodiment within a geographic or mathematical structure.

¤✧¤

The mind's algorithms only suss slivers of Nature. These fractional captures are nonetheless impressive.

> Such structural insights have long been viewed by psychologists and philosophers of science as deeply mysterious in their mechanisms, more magical than computational. ~ Joshua B. Tenenbaum *et al*

Generally, human learning proceeds top-down: getting the big picture first, then using an established framework to fill in gaps. Only when the structural framework is stressed does the mind search for something more robust.

That search is not always successful. Mathematical discoveries have revealed structures to which our mind simply cannot fathom. Science has shown, time and again, that human minds are limited by more than cultural convention (which is quite a constraint unto itself).

Beyond broad characterization, the mechanics of perception and mentation defy our understanding, especially the fluid facility by which learning transpires. To attribute the mind's intricate workings to physical substrates is more mystical incantation than science.

A Mental World

The world is a creation of your consciousness. ~ Nisargadatta Maharaj

The world we experience is entirely within our minds. Our senses are simply a delivery system. We believe what they deliver because the inputs – from sight, sound, and touch – all coincide. That, and others agree with us.

Rationality

Logic is invincible because in order to combat logic it is necessary to use logic. ~ French mathematician Pierre Boutroux

People like to think of themselves as rational. And they often are. But rationality is not the cold calculator of logic that most people think. Rationality is, instead, reasoning that *pleases*.

The dictionary definition of *rational* commonly throws in *sound judgment* as a criterion. Determining sound judgment is an exercise in hindsight; more a metric of success, in which coincidence may play a decisive hand, than of *probability*, which is nothing but a woolly abstraction when it comes to real life.

Beforehand, any risky decision may be considered unsound; and then, in taking a *risk*, one reaches back to probability again. Those in business and technology regularly rely upon risky decisions to propel themselves forward.

All told, *rational* is a vacuous term.

Decisions

Choices are the hinges of destiny. ~ American poet Edwin Markham

Our lives are guided by the decisions we make. Only a relative few of those relate to physical necessities. The great remainder are propelled by desires, most of which have a social seminality.

The mental machinery for decision-making runs a coarse course. Our calculators of logic are shoddy. Stepwise logic is terribly taxing, and cannot consider everything which should weigh in. Hence, we rely upon rough rules of thumb called *heuristics*. There are hundreds of human heuristics – some complementary, some contradictory, most tainted by bias of some sort, all fallible.

Heuristics operate subconsciously. Like a puppet on invisible strings, decisions are mentally presented to us with the built-in biases behind them out of view.

¤✧¤

The mind connects dots of coincidence to construe a pattern, and readily sees patterns where none exist. This is where decisions begin – with available input massaged by the mind.

Here we have the first, and most significant, bias. Our perceptions are heavily filtered by a *framing effect*: viewing situations from a certain perspective, typically personal gain or loss. The context that underlies decisions is itself a bias.

Cognitively, our lives are driven on a bumpy road full of potholes, in hit-and-run mode. Because the lensing is deep within, along with a built-in bias to uphold self-esteem, looking back does not readily provide a corrective view. We tend to see things in a way that justifies what we did. This makes self-correction problematic without questioning the motivations and unexamined assumptions that propel us forward.

First and foremost, decisions are determined emotionally. We live by affect, not reason.

> The mind is always the dupe of the heart. ~ French author François de La Rochefoucauld

As much as we can, we do what we like. By contrast, doing what is good for us – in terms of intrinsic reward aside from the sheer anticipation of satisfaction – falls under the astringent aegis of *discipline*.

> Life is the sum of all your choices. ~ French philosopher Albert Camus

℘ The Mind ☙

Something which I thought I was seeing with my eyes is in fact grasped solely by the faculty of judgment which is in my mind. ~ French mathematician and philosopher René Descartes

The *mind* is an intangible instrument which constantly creates the impression of us having a window onto the world, and into oneself.

Mind and consciousness are not things but processes. ~ American physicist Fritjof Capra & Italian chemist Pier Luigi Luisi

The dream state illustrates that the mind is a fabricator, rather than straight-forward presenter of existence. While awake, what phenomenally appears before us is a phantasmagoric multimedia display, imaginatively sewn from disparate sensations into a consistent story.

The energy of the mind is the essence of life. ~ Aristotle

℘ Conscious & Subconscious ☙

Most of your experiences are unconscious. The conscious ones are very few. You are unaware of the fact because to you only the conscious ones count. ~ Nisargadatta Maharaj

Trickles of thought make their way to one's conscious attention, rising from the deep sea of mentation called the *subconscious*. The subconscious is the engine of cognition. Conscious thought is the fruit from subconscious roots.

℘ Of Two Minds ☙

Only a small fraction of our thoughts do we willfully form and pursue. This intended mentation issues from *willmind*.

In contrast, unbidden thoughts regularly rise to conscious attention from the subconscious; indicating that our minds are largely managed by an agency which is independent of our volition. This is *nattermind*, which is often a nuisance, as it readily distracts with fantasies, doubt, and distress; a deceiver by its nagging worries and schemes.

Nattermind makes the average mind wander away at least one-third of the day. The frequent distraction often lowers mood.

> Although the room seems quiet, it is full of disruptions – ones that come from within. Noisy trains of thought are hard to ignore. ~ Indian economist Sendhil Mullainathan & American behavioral scientist Eldar Shafir

Nattermind's incessancy is troubling for 95% of the population. Most people are discomforted in sitting idle, letting their own minds prey upon them.

> We lack a comfort in just being alone with our thoughts. We're constantly looking to the external world for some sort of entertainment. ~ American psychologist Malia Mason

Nattermind and willmind point out the manifold nature of the mind as receiver, deceiver, and deliverer. The mind is the energetic engine of life: the interpreter of inputs and fabricator of the world; the constructor of concepts, beliefs, hopes and fears; and the tool of all crafts.

Nattermind is not just a nuisance. This cognitive busybody is essential in acting as the gatekeeper of what comes to mind: determining which subconscious stream should surface to awareness.

In the ocean of mentation, nattermind manages the flow of currents. Willmind is but a small craft of volition sailing on the boundless subconscious sea.

In sum: nattermind is, by and large, the mind. Our own willmind is but a narrow peephole.

Schizophrenia

> These are people who are on the extreme end of human experience, who are part of a continuum and not a separate category. ~ American psychiatrist William T. Carpenter

Schizophrenia is a severe illness characterized by mistaking nattermind's fictions for actuality, and thereby often behaving abnormally socially. Common symptoms include false beliefs, confusion, hearing voices (that others do not), reduced sociality and emotional expression.

> Schizophrenia is a modern development. Early hominids did not have this disorder. ~ American psychiatrist John H. Krystal

Schizophrenia typically comes on gradually, beginning in young adulthood, and becomes a chronic condition, albeit with acute episodes.* Schizophrenics often grapple with other mental health problems.

Notwithstanding vast environmental and socioeconomic differences among societies, ~1% of the world's population suffers schizophrenia.

The root of schizophrenia remains unclear to researchers, as the disorder seems to reflect both heritable inclinations and environmental factors. Despite extensive study, geneticists have been unable to zero in on causality, finding individual genes only modestly correlated to the disorder.

> Schizophrenia is so highly, radically polygenic that there may well be nothing to find, just a general, unspecifiable genetic background.† ~ American behavioral geneticist Eric Turkheimer

Environmental stressors seem to play a decisive hand in schizophrenia developing. Epidemiological studies indicate that risk factors range from urban living or being an immigrant to experiencing abuses that include poverty, emotional torment, and sexual predation.

> We need a stronger focus on changing the environment so we can prevent schizophrenia. We need to give children better childhoods and better chances to avoid extreme stress. ~ Norwegian psychologist Roar Fosse

---

* Prior to schizophrenia ever manifesting, children who later suffer the affliction often exhibit flatter emotional states – less joy or distress – and fewer coordinated movements. These early signs may show in children as young as 5 years.

† *Polygenic* means that a plethora of genes are involved.

Though ever-present to those who suffer from it, schizophrenia has no physical cause, and no chemical cure. Instead, a stressed and frenzied nattermind has rudely seized control.

☙ Knowledge ❧

To respect a mystery is to make way for the answer. ~ American poet and philosopher Criss Jami

We all want to understand how the world works – both for our benefit and entertainment. But the seriousness of inquiry varies to quite different extents. Most in the Collective of humanity accept what they are told. Socialization and natural disinclination to academics dulls the desire to delve into the nature of existence.

Nature is often hidden. ~ English scientist Francis Bacon

Only a few seriously seek to uncover Nature's deepest secrets, heading towards what Aristotle called "final cause," which is "that sort of end which is not for the sake of something else, but for whose sake everything else is."

The naturally proper direction of our road is from things better known and clearer to us, to things that are clearer and better known by Nature. ~ Aristotle

☙ Religion, Philosophy, & Science ❧

Beware of false knowledge; it is more dangerous than ignorance. ~ Irish playwright George Bernard Shaw

There are three conspicuous paths to knowing Nature: religion, natural philosophy, and science.

Religion is traditional dogma; a cultural indoctrination of perspective about moral values and the nature of the world. Beyond mythologies of gods and the afterlife, religion is the universal touchstone of human culture – infusing belief systems and infecting other methodologies.

Though they deny being creatures of faith, scientists are inadvertent subscribers to religious doctrines. Such is human nature: to blithely believe while insisting on one's own rationality.

Natural philosophy is the study of nature from a holistic perspective; interpretively accepting what is experienced to build a conceptual realm that constitutes a worldview.

Science descended from natural philosophy, but in doing so, foreclosed the non-empirical with the assumption that manifestation and reality are synonymous.

The current consensus of scientists worldwide is *materialism*: that existence is fundamentally made of matter. This is a vast extrapolation from classical physics theories which explain specific phenomena.

> A strict materialist believes that everything depends on the motion of matter. ~ James Clerk Maxwell

Materialists studiously ignore what physicists learned in the 20th century, because the alternative – immaterialism – is considered unacceptable, as it places reality beyond science's reach.

> Scientists are often accused of a bias toward mechanism or materialism, even though believers in vitalism and in finalism are not lacking among them. Such bias is inherent in the method of science.
>
> The most successful scientific investigation has generally involved treating phenomena as if they were purely materialistic, rejecting any metaphysical hypothesis as long as a physical hypothesis seems possible. The method works. The restriction is necessary because science is confined to physical means of investigation and so it would stultify its own efforts to postulate that its subject is not physical and so not susceptible to its methods. ~ American paleontologist George Gaylord Simpson

Another core belief underlying science is *determinism*: a misplaced sense of surety in causality. Determinism affords a tidy framework for explaining existence as resolving to material cause-and-effect.

> Within science, all causes must be local and instrumental. Purpose is not acceptable as an explanation. Action at a

distance, either in space or time, is forbidden. Especially, teleological influences of final goals upon phenomena are forbidden. ~ English-born American physicist Freeman Dyson

Scientists secondarily embrace a decidedly reductionist bent. *Reductionism* is the belief that complexity can be completely understood by comprehending the constituent components involved.

There is nothing more deceptive than an obvious fact.

~ Arthur Conan Doyle

Faith in reductionism requires that something can never be more than the sum of its parts. Synergy is banished. As this underlying assumption is unreal, reductionism is nothing more than a convenient tool for atomistic study, whilst missing the proverbial forest for the trees.

Just amassing knowledge; there is no point in that. ~ Nisargadatta Maharaj

Dropping the holism inherent in natural philosophy, science moved to the modern perspective by ignoring systemic complexity, and focusing solely on individual components in isolation. While adding vastly to man's fact base, it simultaneously cemented ignorance of reality.

Science may be described as the art of systematic over-simplification. ~ Austrian philosopher Karl Popper

Science's modern methodology is systematically sphinctered to admit only factual evidence, with great favor towards reproducibility. *Empiricism* is the conviction that understanding existence is entirely amenable to human observation. This implicitly assumes full trust in the senses.

Science is not a body of facts. Science is a method for deciding whether what we choose to believe has a basis in the laws of Nature or not. ~ American geophysicist Marcia McNutt

The success that reductionism afforded in explaining a diverse array of static phenomena gave it outsized credence. But reductionism fails in comprehending the nonlinear dynamic systems of which phenomena are invariably composed. For instance, the ecological gyres that characterize life cannot be accounted for under reductionism, which can only go as far as chemical reactions.

> Science is a system of idealised entities: atoms, electric charges, mass, energy and the like – fictions compounded out of observed uniformities, deliberately adapted to mathematical treatment that enable men to identify some of the furniture of the universe, and to predict and control parts of it. ~ French philosopher Georges Sorel

The enormous irony of empirical science is that its power is firmly grounded in abstraction, and the imagination. For all logical reasoning involves *counterfactual thinking* for its assessment – the ability to imagine what does not exist.

> Innocent, unbiased observation is a myth. ~ Brazilian-born British biologist Peter Medawar

Science often stumbles into two major pitfalls: making unwarranted assumptions, and fallaciously attributing causality.

An *assumption* is taking something for granted, a supposition. In an open inquiry about reality, nothing should be assumed.

> Sit down before a fact as a little child, be prepared to give up every preconceived notion, follow humbly and to whatever abysses Nature holds, or you shall learn nothing. ~ English biologist Thomas Henry Huxley

The other common error in science is mistaking correlation with causality. This readily arises from the innate way our minds work.

There is a natural propensity to *induction* – reaching broad conclusions from a small pool of observations – that both drives us towards deeper truths, and, through a desire for simplification, forms the basis of false beliefs.

Astute observers sense that there is more going on than first appearances convey. So, there has been a restless search for general principles, and for the foundations underlying existence. This was the thrust of natural philosophy that was inadvertently abandoned by science in its embrace of material empiricism.

Therein sits an open secret. Science has provided ample evidence which may be drawn upon to understand Nature and its derivation. But refusal to step beyond the phenomenal blocks the way. Behind this obstinacy is the religious belief that actuality is reality. Ignorance is learned, and then insisted upon. Hence the myth of *scientific materialism*: that reality is ultimately physical. It is a naïve empiricism.

> It is far better to grasp the Universe as it really is than to persist in delusion, however satisfying and reassuring. ~ American astronomer Carl Sagan

Laws of Nature

> To be a scientist, you have to have faith that the universe is governed by dependable, immutable, absolute, universal, mathematical laws of an unspecified origin. ~ Paul Davies

Science attempts to render the order we perceive ironclad by formulating laws of Nature. This distillation invariably ends in symbolism: mathematics.

> It was mathematics, the non-empirical science par excellence, wherein the mind appears to play only with itself, that turned out to be the science of sciences, delivering the key to those laws of Nature and the universe that are concealed by appearances. ~ American theorist Hannah Arendt

Mathematics

> Mathematics is the gate and key of the sciences. ~ English philosopher Roger Bacon

Via mathematics, science aims at discovering the laws by which existence is encoded. Physics models pervasively show extra-dimensionality, and entanglements involving infinities; indicating phenomena beyond our ken. As a compass to the truth, mathematics points to existence as a practically inscrutable complexity. Biology seconds this apprehension.

> Why Nature is mathematical is a mystery. The fact that there are rules at all is a kind of miracle. ~ American physicist Richard Feynman

❀❀❀

The real beauty of life is in orderliness. ~ Ghanaian writer Ernest Agyemang Yeboah

Every localized field, every object, and every interaction that ever exists or transpires is unique. But our minds have an inexorable inclination to categorize – to view the world through templates. By this tendency we create an order to Nature, which is different than the order Nature creates.

We gain our ends only with the laws of Nature; we control her only by understanding her laws. ~ Polish-born British mathematician Jacob Bronowski

Most importantly, we do not readily see the world as a series of interacting processes. Time is a secondary sense. Hence temporal truths – of consequences – are the hardest lessons learned. Instead, strong spatial orientation – towards objects and bodies – issues its own delusional consequences.

Forgive the barbarian, for he believes that the customs of his tribe are the laws of Nature. ~ George Bernard Shaw

A counterforce to complacency compels. Drawn like a moth to flame, there is a natural impulse to seek seminal causes. Following convention, the miracle of Nature having laws behind its exhibition invoked a Creator.

The laws of nature are but the mathematical thoughts of God. ~ Greek mathematician Euclid in the 4th century BCE

The urge towards theism is a strong statement of an imperative need to sense a cosmic order, and doing so through objectification. But an equally poignant declaration was made in believing that Nature has laws.

Whereas natural law remained a central precept, post-industrial science gave God a special dispensation: banishment. Scientists still have faith. But it has been firmly placed in a natural order without a supernatural cause.

The ability to reduce everything to simple fundamental laws does not imply the ability to start from those laws and reconstruct the universe. ~ American physicist Philip W. Anderson

Like energy that exudes materiality, the only things propping up laws of Nature are abstractions, bridled to our need to render order. Sensing structure makes us feel safe, and offers the profitable prospect of predictable exploitation. So, we celebrate such laws, and consider them real.

> The great delusion of modernity is that the laws of Nature explain the universe for us. The laws of Nature describe the universe, they describe the regularities. But they explain nothing. ~ Austrian philosopher Ludwig Wittgenstein

꧁ The Limits of Knowledge ꧂

> There's something wrong with the world. You don't know what it is. But it's there. ~ Morpheus, in the movie *The Matrix* (1977)

Intellectual fallibility is intrinsic in the life of any sentient being. The problem is existence itself.

> We may regard the present state of the universe as the effect of its past and the cause of its future. An intellect which at a certain moment would know all forces that set Nature in motion, and all positions of all items of which Nature is composed, if this intellect were also vast enough to submit these data to analysis, it would embrace in a single formula the movements of the greatest bodies of the universe and those of the tiniest atom; for such an intellect nothing would be uncertain and the future just like the past would be present before its eyes. ~ Pierre-Simon Laplace in 1814

Pierre-Simon Laplace was one of the greatest scientists of his time. Peculiarly, Laplace's work on probability led him to certainty – a conclusion which became known as *Laplace's demon*: that omniscience would render existence entirely predictable for all time. This was the first published articulation of *scientific determinism*.

In crowning causality as the sovereign of science, the deterministic demon proved inspirational to the scientifically minded that followed in Laplace's wake. But it was a sophistic siren.

Deep in the deluge of knowledge that poured forth in the 20th century were ironclad limits on what can be known.

In 1926, addressing what might be learned about the qualities of the fundamental quanta of physicality, German theoretical physicist Werner Heisenberg discovered inherent uncertainty. The primordial constituents of matter were inescapably probabilistic at best. Since then, Heisenberg's uncertainty principle has been proven an unshakable pillar of certitude.

In 1931, Austrian-American logician Kurt Gödel argued that, for any formal mathematical system to be useful, it is impossible to use the system to prove every truism it contains. Mathematically speaking, truth involves faith at some point. Gödel's incompleteness theorems are indispellable.

In 2008, American mathematician David Wolpert logically proved that all knowledge of any possible universe is beyond the grasp of any intellect that might exist within it. So much for omniscience.

Wolpert put the last nail in the coffin of Laplace's demon. But he did more than that. In determining the "physical limits of inference," Wolpert showed that there are facts about phenomena which cannot be phenomenally known, by either experiment or computational prediction. Hence empiricism can never unravel existence, and so cannot provide a correct scientific worldview (Weltanschauung).

Truth is exact correspondence with reality. ~ Indian yogi and guru Paramahansa Yogananda

ꕤ Falsity ꕤ

History shows again and again, how Nature points out the folly of men. ~ American musician Donald Brian Roeser, in the song "Godzilla" (1977), performed by Blue Öyster Cult

A cold-eyed look at humanity cannot help but bring chagrin. History has been a whirlwind of incomprehension that continues to this day.

The quintessential question is *why*? Why the continuing folly of men? The root answer is simple: whereas discerning actuality takes effort, believing misinformation is easy.

One can never reverse what has been learned, even if the uptake is fiction, not fact. The mind refuses to return to a prior state.

> It is extremely difficult to return the beliefs of people who have been exposed to misinformation to a baseline similar to those of people who were never exposed to it. ~ Australian psychologist Stephan Lewandowsky *et al*

The inability to subtract what is known is called the *curse of knowledge*. Its common expression occurs when teachers or writers assume their respective students or readers know more about a subject than they do.

> Social interaction is really an interaction of minds, of mental states. ~ Janet Wilde Astington

Culture relies upon sharing information. Although believability is a factor in determining whether information is propagated, people mainly pass on tidbits that evoke an emotional response, irrespective of veracity.

> Emotional arousal increases people's willingness to pass on information. ~ American psychologist Colleen M. Seifert *et al*

Social repetition creates an illusion of consensus when none exists. Thinking that others believe something – particularly those who are respected – solidifies and sustains belief in a falsity.

> A lie told often enough becomes the truth. ~ Russian revolutionary Vladimir Lenin

From ancient times, propagandists have appreciated human gullibility. Politicians invariably rely upon it.

One outcome of lies in the societal arena is *pluralistic ignorance*: a dichotomy between what is generally believed and what people think is generally believed. This can result in the *false-consensus effect*: those with a minority belief wrongly thinking that they are in the majority.

Such wrongly perceived social consensus can serve to solidify belief in falsity. Correction becomes practically impossible; creating what is known as the *continued influence effect*. A lie becomes the truth.

The press contributes to misinformation in a few ways. To render science "newsworthy," the media inexorably oversimplify, misrepresent, or overdramatize scientific results.

As science is inherently complex, simplification is inevitable. But oversimplification easily leads to misunderstanding.

Another failure prances in journalists' illusory aim to be "balanced" in their reportage. There is no balance to be had in some stories.

> If media stick to journalistic principles of "balance" even when it is not warranted, the outcome can be highly misleading. ~ Australian psychologist John Cook *et al*

Reportage about manmade climate change is exemplary. Human manufacture of global warming through worldwide pollution is scientifically indisputable. Yet contrarian tripe has featured prominently in the media. The misleading sense of controversy on a scientifically-settled issue has left the public confused about the severity of human impact on the planet.

While refreshingly diverse, the cacophony of the Internet has grossly exacerbated the problem of public misinformation. So-called "fake" news has become common.

Fact or fiction, repetition in of itself reaps a reward when it comes to learning. Familiarity creates credence.

> Repetition of information strengthens that information in memory and thus strengthens belief in it, simply because the repeated information seems more familiar or is associated with different contexts that can serve as later retrieval cues. ~ Australian psychologist Ullrich K.H. Ecker *et al*

Falsity is most readily accepted when it is part of a plausible story, consistent with other assumed facts, or corresponds with established beliefs.

> People's worldview plays a key role in the persistence of misinformation. Personal beliefs can facilitate the acquisition of attitude-consonant misinformation, increase reliance on misinformation, and inoculate against the correction of false beliefs. ~ German-American psychologist Norbert Schwarz *et al*

Only if a source is considered credible, and the listener relatively open-minded on a topic, can otherwise suspicious factual content hope to trump its sinister sister, misinformation.

> Although suspension of belief is possible, it seems to require a high degree of attention, considerable implausibility of the message, or high levels of distrust at the time the message is received. So, in most situations, the deck is stacked in favor of accepting information rather than rejecting it, provided there are no salient markers that call the speaker's intention into question. Going beyond this default of acceptance requires additional motivation and cognitive resources. ~ Stephan Lewandowsky *et al*

Once misinformation is accepted, it is highly resistant to deletion, which requires active mental denunciation. This is especially difficult when a falsity fits well within an ensconced context.

> Retractions rarely, if ever, have the intended effect of eliminating reliance on misinformation, even when people believe, understand, and later remember the retraction. ~ Norbert Schwarz *et al*

Misinformation corresponds with, and is abetted by, the *illusion of knowledge*: people thinking they know more than they do. With few exceptions, the illusion of knowledge is universal.

> The greatest obstacle to discovery is not ignorance – it is the illusion of knowledge. ~ American historian Daniel J. Boorstin

Thus, this is the state of human knowledge: an admixture of facts and fabrications attached to widespread beliefs in falsehoods, with fictions as foundations, and facts supportive only by being sewn into a fabric of falsity. The crucial beliefs of humanity are rubbish which have diverted humanity down cul-de-sacs of self-destruction.

> The bulk of the world's knowledge is an imaginary construction. ~ American author Helen Keller

Beyond Phenomena

What we call reality in actuality is our perception of it. ~ American singer-songwriter Jewel Kilcher

We take for granted that what we perceive *is* reality. Belief in our senses is instead a fantastic cage we build for ourselves, and so bound our comprehension.

The senses deceive from time to time, and it is prudent never to trust wholly those who have deceived us even once. ~ René Descartes

The Nature of Reality

I think, therefore I am. ~ René Descartes

Though we know thought to be abstract, science insists that the brain conjures the mind. At issue here is the heart of all philosophic thought and scientific endeavor – the nature of existence, and, by extension, the essence of reality.

Three possibilities exist: dualism, materialism, and immaterialism. Our everyday experience is *dualism*, with a distinct body and mind encountering an outside world.

The world is seen only due to duality. If there is no duality, there is no world. ~ Nisargadatta Maharaj

If, instead, reality is a unity, the proximate illusion of duality flows from a singular source. The wellspring of existence may either be from matter (materialism); or otherwise involve a unified Consciousness (immaterialism), which somehow creates a figment of materiality. Materialism and immaterialism are diametric concepts. If existence emanates from a unicity, it must exclusively be either material or ethereal.

Dualism

Dualism is the traditional take; the belief that existence is somehow bifurcated between body and mind, with an interface between the two. Therein is the problem of dualism.

Throughout history, despite relentless investigation, not even a scintilla of evidence for an interface between body and mind has ever been found. The *mind-body problem* has proved unsolvable.

> Even though everybody agrees that mind has something to do with the brain, there is still no general agreement on the exact nature of this relationship. ~ Finnish psychologists Antti Revonsuo and Matti Kamppinen

The root issue of the mind-body problem is trying to entangle a duality; how do two very different phenomena intimately interrelate?

> If this dualistic theory were true, it would confront us with the most embarrassing, insoluble difficulties should we try to explain how these two utterly different substances could interact with one another, as they appear to do in human behavior. ~ American philosopher Mortimer J. Adler

If duality is an illusion, the mind-body problem dissolves. Either the brain manufactures the mind, or the mind fabricates a material mirage.

℘ Materialism ☙

With dualism downed, reality must instead be a monism. Scientists tout *materialism*: the metaphysical belief that measurable matter and energy are fundamental.

> Materialism is the view that only the physical world is irreducibly real, and that a place must be found in it for mind, if there is such a thing. This would continue the onward march of physical science, through molecular biology, to full closure by swallowing up the mind in the objective physical reality from which it was initially excluded. The assumption is that physics is philosophically unproblematic, and the main target of opposition is Descartes' dualist picture of the ghost in the machine. ~ American philosopher Thomas Nagel

The focus of materialism is matter. That energy exists is undeniable, but materialists make no attempt to account for energy beyond treating it as measurable.

> The ontology of materialism rested upon the illusion that the kind of existence, the direct 'actuality' of the world around us, can be extrapolated into the atomic range. This extrapolation, however, is impossible. Atoms are not things. ~ Werner Heisenberg

Most saliently, materialism posits that consciousness and mentation are figments of the brain. Consciousness, coupled with the workings the mind, is thought to come from cellular reactions of chemistry and electricity.

> Our own consciousness is a product of our brains. ~ Canadian cognitive psychologist Steven Pinker, firmly planted in the mire of materialism

The grand delusion of materialism comes in confusing correlation with causality. Physical intelligence system activities, such as brain waves and chemical reactions, may synchronously coincide with mentation. But it's not the brain telling the mind what to think.

♎ Mistakes ♎

Consider mistakes. How well people bounce back from blunders depends upon what they believe about the nature of intelligence.

Those who believe that savvy develops through effort view their flubs as learning opportunities. But to those who hold that intelligence is fixed, mistakes indicate lack of ability.

The mental and physical response to mistakes differs between these two groups. Someone in the growth mindset becomes more attentive after making a mistake, and so their accuracy improves. This betterment does not occur in a person with a fixed mindset.

> A growth mindset is associated with heightened awareness and attention to errors as early as 200 ms following error commission. ~ American psychologist Jason S. Moser *et al*

Moser and his colleagues found this out by experimentally measuring electrical activity in the brain of participants in tests designed to provoke errors, and later asking the participants their beliefs about intelligence.

> Larger amplitudes of event-related potentials – electrical brain signals elicited by events – are associated with adaptive behavioral adjustments, such as slower and more accurate responses following mistakes. ~ Jason S. Moser *et al*

> Nobody has the slightest idea how anything material could be conscious. Nobody even knows what it would be like to have the slightest idea how anything material could be conscious. ~ American cognitive scientist Jerry Fodor

It is impossible to construe how the brain could create beliefs. But it is entirely reasonable, backed by extensive evidence, to think that the mind can affect the physical body, including the brain.

The effective potency of placebos is one of innumerable known examples of mind-over-matter which materialism cannot account for. Another broad area of immateriality at work involves the well-known deleterious effects that mental stresses have on the body. Whereas mental attitude can be curative, the brain has no such power.

♎ Caring & Control ♎

Babies are motivated by the biological desire to find out how to effectively deal with their environment. From a tender age to one's dying breath, people strive for control.

Most people gain greater control over their environment as they grow older. That trend is often abruptly reversed when infirmity strikes the elderly. Many aged people experience further decline after being institutionalized.

> The consequences of such a loss of control usually include withdrawal, depression, and sometimes early death. Experimental studies with animals, including humans, amply demonstrate the negative effects of loss of control. ~ American psychologist and gerontologist Richard Schulz

In the mid-1970s, Richard Schulz conducted an experiment on control among the institutionalized aged. A group of his undergraduate students visited old folks in a retirement home for two months. The visitors were strangers to the elderly people visited.

For the study, retirement home residents were divided into four groups. One group received no visits. A second group was visited at random. A third group was told when the visits would occur, and how long they would last, but otherwise they had no control over the visits.

The fourth group was given complete control of the visitation. They could decide not only how long each visit lasted, but how often they were visited.

Residents who were most able to predict and control the visits became happier, healthier, and more hopeful than other residents. Their new-found zest for life meant more activity, less boredom, and fewer medications. They looked and acted less like old people. Schulz concluded that controllable visitation "actually reversed the pattern of progressive decline."

The coda to this story is a sad one. Follow-on study showed that gloom again descended after the visits ended.

The toll was especially high for those in the group able to regulate their visits. For them, losing the sense of control that had enlivened them was deadening. Their zest for living evaporated, and their health precipitously declined.

⌑✧⌑

> Having control over a stimulus means that it is predictable. It becomes important to ask, therefore, whether the ability to control adds something over and above the ability to predict.
> ~ Richard Schulz

Schulz's retirement home study showed that exerting control creates the ineffable product of inestimable value: joy. The jubilant bounty of living comes from exercising meaningful will.

> Total care for the aged is just as bad as no care at all. Other animal studies have demonstrated that organisms prefer working for positive reinforcement over securing them for free.
> ~ Richard Schulz

❀❀❀

Beyond the powers of the human mind, materialism has no explanation for how organisms without identifiable brains could possibly behave intelligently, from microbes on up. The savvy of plants is indisputable; yet they have no physical system for cognition.

¤✧¤

That the mind and body are entangled is obvious. But the brain is not causal; nor, for that matter, is the mind, which plays its part as presenter of physicality via symbolic processing.

Materialism ignores modern physics: that matter is energy transposed, and that energy is a fabrication of Nature, but only a chimera by which existence manifests. And *adaptation* – the irrefutable momentum behind biological evolution – cannot be explained via materialism.*

Albeit convincing by appearance, materialism is ultimately unreal. Scientifically, it leaves far too much mystery.

> A new scientific truth does not triumph by convincing its opponents and making them see the light, but rather because its opponents eventually die, and a new generation grows up that is familiar with it. ~ Max Planck

℘ Immaterialism ℘

> It all looks fine to the naked eye, but it don't really happen that way at all. ~ English musician Pete Townshend, in the song "Naked Eye" (1971)

Because it seems so outlandish (as does quantum physics, for that matter), few apprehend *immaterialism* as the nature of reality. Under immaterialism, consciousness passively witnesses while the mind cognizes, with a physical intelligence system as a relative correlate. The mind and body are an apparent, but illusory, duality; ultimately formed of energy, which is immaterial.

---

* Note that there is a difference between the mechanics by which adaptation appears to work, and the goal-directed process (teleology) by which it proceeds. Materialism can partly account for evolutionary mechanics via genetics, but not the impetus behind adaptation.

Appreciating immaterialism scientifically requires careful recognition of the difference between proximate appearance and ultimate reality. A similar issue arises in quantum physics: is a quantum a particle or a coherent, localized field? The answer is an interaction, not an object; a field, not a particle. But all that ever may be experienced of a quantum is in its apparent materiality. Fields are only felt from their effect on matter.

Before proceeding, we must come to terms with this strange landscape of a fundamental immateriality. Under immaterialism, what we call *existence* is only proximate. Reality is ultimately of *noumenon*: nonexistence.

In that the mind's workings can be observed (introspection), the mind cannot be considered the first cause. The witnessing capability of awareness – consciousness – is passive.

Actuality and reality are day and night. There is a vast, irreducible void between proximate phenomena and an ultimate noumenon. The bustle of multifarious existence emanates a singularity of conscious nonexistence.

The tangible world is a mirage. The brain putters away, but it and other material objects are ultimately figments of the mind, without independent existence.

The world exists within oneself. ~ Nisargadatta Maharaj

Energy

Energy is more a 'scientific' idea than substance. ~ English chemist and molecular biologist Graham Cairns-Smith

Energy is the coherent medium by which existence manifests. But construing energy as confined within any physical system known as the universe is wrong, because energy is immaterial. The laws of thermodynamics are fictional.

Energy is what it takes to get matter to work. Only the effects of energy on matter are phenomenal. Any measurement of energy is always the impact that it has on a material substance.

Understanding energy for what it is (an abstraction), and what it is not (material), you can see that the seemingly unbounded energy of the ground state is mere mathematics. If the nothingness of vacuum has energy beyond imagination, how can it be phenomenal?

The gravitational force of so-called "dark matter" creates entropic distortions in the fabric of the universe. Dark matter is nothing more than energy spicing spacetime with a fractal cosmic pattern that guides celestial affairs; a stringy equivalent to black holes, which steer galaxies. Along with the ground state, these are energetic instances of nothing coherently shaping everything.

Immaterialism has been scientifically validated innumerable times. This is universally ignored because researchers never question their assumption of materialism. Instead, scientists are sometimes left scratching their heads at inexplicable discoveries. Conversely, some who should be brain dead aren't.

♎ Alzheimer's Disease ♎

> What's surprising is these people exist. ~ American cognitive neurologist Changiz Geula

In 1906, German physician Alois Alzheimer had a patient with sudden, profound memory loss, among other psychological deteriorations. In an autopsy, he found her brain physically ravaged. How she had managed before her memory crisis was inscrutable.

An unknown percentage of people with brains chock-full of the plaques and tangles that characterize Alzheimer's disease maintain razor-sharp memory and mental acumen.

> These people, for all practical purposes, should be demented. ~ Changiz Geula

This is but one of innumerable scientific findings which cannot be accounted for by materialism.

♎ Vision ♎

Even the simplest visual scene is nothing short of a miracle.
~ Indian neuroscientists Chaipat Chunharas & Vilayanur S. Ramachandran

Our sight seems so effortless that we take it for granted. But vision is physiologically impossible, for numerous reasons. Here are just a few.

First is the nature of photons. Sight is matter of refractive absorption and translation. We can see things because the atoms of objects from which photons bounce off absorb particular photonic wavelengths. Every time a photon is banged up, its story changes.

The human eye is an odd construction. Before being absorbed by the cells that receive the message which light has to offer, photons must first pass through the nerve tissues that act as signal transmitters to the brain. The cells that absorb light are at the back of the retina.

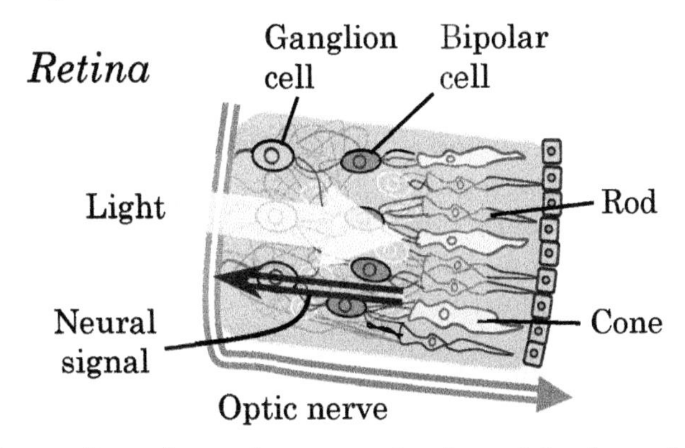

Given that photonic energy is altered by its substantive encounters – the very nature by which vision works – how the light patterns that strike the eyeball make their way to the back of the retina unmolested is inexplicable.

Then there is the problem of translating the photonic wavelength data into perfectly accurate information that the brain can use. Neurons are not optic cables. They use both chemical and electrical signaling. According to materialist dogma, precise light energy compositions are perfectly transcribed multiple times before the brain recreates the exact imagery of the outside world, which came in through the eye upside-down and backwards.

> Your only high-resolution vision is in the very center of your eye – about 0.1% of your entire visual field. ~ American neuroscientists Stephen L. Macknik & Susana Martinez-Conde

Then there is the problem of arranging the billions of photons received into a well-focused panoramic picture fast enough. What we see as the outside world is a massive montage, inscrutably assembled. Tens of thousands of snapshots must be collated, correlated, and exactly aligned into a seamless image in less than 1/100th of a second.

> Information that the brain receives from the eye is already out of date by the time it gets to the visual cortex. ~ Gerrit Maus

Then there's blinking. Humans blink their eyes about every 4 seconds. A blink lasts one-tenth of a second. And every time it happens, the eyeball moves slightly.

Yet visual imagery remains constant despite frequent outages and misdirection in eye movement. This cannot be accomplished through purely physiological means. Research has shown that there must be image consistency at least every 13 milliseconds not to detect jitter; far faster than the time it takes the eyes to blink.

♎ Sleepwalking ♎

> A dissociate state of arousal may modify the components of sleep-wake behavior, consciousness, and also pain perception. ~ French psychiatrist Régis Lopez

Sleepwalkers present an intriguing paradox – although they are prone to headaches while awake, whilst sleepwalking they are unlikely to feel pain, even upon serious physical injury. There are many instances of sleepwalkers falling and breaking bones without waking up.

It is hard to credit this happening if the brain were running the show. Only if an energy system manufactures the mind and physicality does this make any sense.

♎ The Cellular Effects of Meditation ♎

Meditation can influence key aspects of your biology. ~ American physician Linda E. Carlson

Telomeres are the protein complexes at the end of chromosomes that protect chromosomal integrity. Telomere length is critical to health.

Shortened telomeres are associated with cell aging and several disease states. Longer telomeres are thought to provide some disease protection.

After just three months, the telomere length of breast cancer survivors who took up meditation was maintained, while those who did not meditate suffered shortened telomeres. Cells respond to meditation in several ways, including improved genetic expression.

The calmness of our mind can influence gene expression. ~ American psychologist Richard J. Davidson

☙ Consciousness ❧

A scientific world-view which does not profoundly come to terms with the problem of conscious minds can have no serious pretensions of completeness. Consciousness is part of our Universe. So any physical theory which makes no proper place for it falls fundamentally short of providing a genuine description of the world. ~ English mathematical physicist Roger Penrose

The obvious stumbling block to immaterialism is how individual minds might share the same experience of duality. The answer is simple, if fantastic: each individual consciousness is a particulate manifestation of a unified, universal field of Consciousness.

Consciousness itself is the source of everything. ~ Nisargadatta Maharaj

To populate the performance of phenomena, Consciousness infuses life with awareness.* It is not that individuals each have their own consciousness. Rather, Consciousness assumes innumerable forms.

> Every organic being, every autopoietic cell is conscious.
> ~ American evolutionary theorist Lynn Margulis

Consciousness pervading life at all levels is subtly apparent by the fact that not only do organisms possess awareness and proceed with purpose, but so do cells within a body, the organelles of cells, and the workers within: proteins. The discretionary decisions involved in intracellular coordination, adaptive immune systems, the healing of wounds, and intercellular communication are examples where physical forces alone cannot explain the intelligence behind the processes.

As with quantum particles, objects and bodies are artifacts of coherent energy fields. Nature is nothing but an ongoing process. Our everyday experience of individuality and separateness is make-believe.

Consciousness is the passive platform for awareness. Its active counterpart – Nature – nudges with discerning deportment.

ꝏ Coherence ꝏ

> Whence arises all that order and beauty we see in the world?
> ~ Isaac Newton

The interactive force behind Nature is *coherence.* In other words, coherence is the mind of Nature. That Nature has order of seemingly infinite complexity is the testament of coherence.

> In all chaos there is a cosmos, in all disorder a secret order.
> ~ Swiss psychiatrist and psychotherapist Carl Jung

---

* Consciousness (with a capital 'C') is a unified, non-phenomenal field which manifests in particulate form (*consciousness* with a small 'c'), bound to individual mind-bodies.

Classical physicists identified four natural forces: light, electricity, magnetism and gravity. Modern physicists combined the first three, and determined that gravity is an entropic warpage of spacetime caused by concentrated mass; quite dissimilar from energetically active electromagnetism. Quantum physicists identified two nuclear forces: strong and weak, which respectively hold atomic nuclei together and have them decay.*

Physicists conjecture that all forces are unified at some level, as they were at the surmised instant that was the birth of our universe, when the cosmos was a supposed singularity. Their suspicions are correct.

Coherence is the first, fundamental force, with infinite energy at its disposal. Hence the infinities that appear in mathematical descriptions of quantum mechanics and geometric relativity. Energy is the immaterial by which coherence weaves materiality.

To date, physicists have sidestepped the most basic issue about existence. They have had nothing to say about the font from which energy flows.

Instead, quantum physicists confine themselves to marveling at the complex coherence of patterns by which existence is formulated bottom-up.† Meanwhile, astrophysicists ponder the power by which star systems, galaxies, and even larger cosmic structures cohere in precise interplay.

Physicists puzzle over patterns which may be discerned mathematically. Sometimes such patterns converge in widely divergent phenomena, illustrating a fundamental algorithm at work.

---

* The term *interaction* is now preferred by physicists over the brusquer *force*, as interaction emphasizes that energy only changes via interactivity. Existence is ever ecological.

† The starting point is, as a vexed Max Planck observed, Nature's insistence on quantizing waveforms as the fundamental basis for materiality. The *Planck constant* is the measure of that obstinacy.

♎ Cells & Neutron Stars ♎

> Seeing very similar shapes in such strikingly different systems suggests that the energy of a system may depend on its shape in a simple and universal way. ~ American nuclear physicist Martin Savage

The endoplasmic reticulum (ER) is an organelle in eukaryotic cells comprised of stacked sheets connected by helical ramps. The stacked arrangement affords maximum workspace in a minimal amount of physical space. The ER has a myriad of functions critical to cell vitality, including protein assembly.

A neutron star is the collapsed core of large star. Neutron stars are the smallest and densest stars known. These exhausted stars are 14 orders of magnitude denser than the environment found in living cells.

The size difference between ER and neutron stars is a million orders of magnitude. Yet they share the same basic structure.

The crust of neutron stars comprises layers of dense nuclear matter connected by ramps. This nuclear pasta has a selfsame arrangement as endoplasmic reticulum.

> For neutron stars, the strong nuclear force and the electromagnetic force create what is fundamentally a quantum-mechanical problem. In the interior of cells, the forces that hold together membranes are fundamentally entropic and have to do with the minimization of the overall free energy of the system. At first glance, these couldn't be more different. ~ American condensed-matter physicist Greg Huber

In that an order is apparent in existence, there must be a composer of it: coherence.* That our minds are inclined to

---

* Note the word "composer" implies a being, not a process. This is a flaw of language, which is strongly object-oriented. Speaking in terms of process is more convoluted than talking about objects. Hence the only economical course is to refer to processes as if they were objects. This unfortunate phrasing turns up several times herein.

view the beauty of Nature as organized patterns reflects how individualized consciousness corresponds with unified Consciousness.

The symphony of patterned existence is written in mathematics, which is the language of Nature.

> Reality is a mathematical structure. ~Swedish-American cosmologist Max Tegmark

While physics provides an apt landscape upon which coherence comes into view, the most cogent argument for the unity of coherence and Consciousness is life.

Consciousness and coherence are described as if they are independent channels. They are not.

Monism means unity. Any duality, or multiplicity, that seems to arise is ultimately fictive. As such, existence is a simulation of physicality, arising from the mind as entertainment.

Our minds have a natural tendency towards *factor analysis*: to tear apart complexities into simpler units to facilitate comprehension. Along with this is the inclination towards reductionism, which is another analytic simplification technique.

Because of these proclivities, we have a hard time grasping the notion of *synergy*: that the interaction of constituent elements produces a greater effect than the individual elements can. Holistic understanding is hard to come by.

The rampant environmental destruction that humans have unleashed upon Earth owes in large part to our inability to comprehend the importance of healthy ecosystems as holistic processes. We instead think of individual animals and plants, and their population numbers. The extinction event now underway fundamentally owes to erroneous abstractions, welded into erroneous belief systems.

Such is economics – a conceptual reductionism which has senselessly left the majorities of peoples around the world struggling to survive, while a tiny minority wallow in wealth.

Capitalism, which eschews large-scale cooperation for minor league contention, is irrational in many ways. Yet the market system dominates humanity because its core atomic abstraction is appealing: wealth attained by competition; winners and losers. That, and people don't care enough about each other.

ഇ Existence ര

The phenomenal world is ever-emergent – continuously created. The only moment that manifests is the ever-present *now*. The future is a projection, and the past a memento of memory. Both are products of the mind.

But so too is the present. Our experience of each instant is a fabrication gleaned from disparate inputs: many millions of photons processed into a pastiche seen as *vision*; rumbling in an aural range of frequencies sounded out as *hearing*; chemical vibrations that come off as *smell* and *taste*; and the sensate sense of proximity termed *touch*, forged from harmonic interactions at cell membranes.

How do you know when you touch something wet? There are no skin receptors for sensing moisture.

Wetness is a perceptual illusion: a mental construction through intricate multisensory integration. We learn to perceive moisture through repeated encounters, soaked in the effects.

ഇ Requisites ര

As an entertainment platform, existence has three essentials: diversity, continuity, and entanglement.

A universe of singularity would be practically nothing: energy without frequency, which is unimaginable. Hence diversity, which defines every facet of existence.

Every identifiable thing – inanimate or alive – is unique. Mental categorization is merely a convenience for managing potential interaction.

♎ Amazon Mollies ♎

> Behavioural individuality is thought to be caused by differences in genes and/or environmental conditions. Therefore, if these sources of variation are removed, individuals are predicted to develop similar phenotypes lacking repeatable individual variation. ~ ichthyologist David Bierbach *et al*

The Amazon molly is a small freshwater fish, native to Mexico and southeast Texas. The term *Amazon* in the molly's name refers to the female-run society in Greek mythology. For Amazon mollies are an all-female species, reproducing clonal daughters.

A "tightly controlled" experiment tested "whether near-identical rearing conditions and lack of social contact dampen individuality," as predicted. Instead, each molly showed her own unique personality.

> Substantial individual variation in behaviour emerges among genetically identical individuals isolated directly after birth into highly standardized environments. ~ David Bierbach *et al*

Without continuity there could be no comprehension, as there would be no consistency or predictability. Whence arises the vector of time.

For individual life forms, continuity provides a platform for memory, and its temporally-vectored converse: the future – which provides for goals, problem-solving and planning. In short, something worth living for.

The dimensions of space provide localization, which, coupled to linear temporality, afford a venue for both dynamic expression and diversity. Spacetime is the outcome of the requirements for diversity and continuity.

Coherence achieves complexity by combining diversity and entanglement. But entanglement serves a much more fundamental purpose: it is the glue of diversity, facilitating ecological intricacy among individuals. The field that delimits every object and life form is woven with entanglement.

Life is a tangled diversity at every level. Populated by legions of proteins working together, even the simplest cell is brimming with intertwined complexity. Multicellularity exponentially entwines synchrony with variety to achieve macroscopic life.

Entanglement is apparent at every scale; as are diversity and complexity. Entanglement is an acknowledged quantum phenomenon, but is also plays a vital role at the ambient scale. Ecosystems are an expression of entanglement. Cosmological construction is driven by black holes – massive nothingness entangled with everything.

Beauty

The sublime *moves*, the beautiful *charms*. The sublime must always be great; the beautiful can also be small. The sublime must be simple; the beautiful can be adorned and ornamented. ~ German philosopher Immanuel Kant

Aesthetics is an essential aspect of Nature as an entertainment platform. Among animals where mate selection is the province of the discriminating feminine, females choose mates with features that indicate physical fitness or skill at foraging. Shapes go wonky and colors go flat if a male is not immunologically buff. Antlers need lots of calcium, and the bowers of birds unstinting dedication.

In other instances, an evolutionary handicap principle applies. It is hard to stay alive adorned with a huge or brilliantly colored attraction. Such a male that lives to show off must be sporting a genetic package worth carrying into the next generation. Sometimes natural beauty illustrates pure aesthetic evolution – that beauty is itself enough.

The most refined beauty may serve as a sexual charm, and for no other purpose. ~ Charles Darwin

⌑✧⌑

Sensuous feeling can take place without any thought whatever. ~ Immanuel Kant

Kant claimed that experiencing beauty was a thoughtful endeavor, but that sensuous pleasures, such as eating and sex, were not. Kant was half right.

Beauty is strongly pleasurable, and strong pleasure is always beautiful. Both require thought. ~ American psychologist Denis G. Pelli

The mind is ever engaged in experiencing the world, even when doing so indolently. Here we have another primary principle: life is an exercise in cognition, in a universe of beauty.

Graciousness is the beauty of virtue. ~ Immanuel Kant

~ Enlightenment ~

The knowledge of an ultimate immateriality, and a unified field of Consciousness, has been known for millennia by a relative few who became *enlightened*. The conveyance of this esoteric truth has been understandably misunderstood.

In the middle of the first millennium BCE, Chinese sage Lao Tzu wrote of the *Tao*: the absolute stillness that is the source of Nature.

The great Tao extends everywhere. It does not have a name. The nameless originated Heaven and Earth. Tao acts through natural law. All things depend upon it for growth, and it does not deny them. Evolved individuals hold to the Tao, and regard the world as their pattern. ~ Lao Tzu

In 3rd century BCE, legendary Indian guru Patañjali instructed the path to enlightenment through meditation:

Freedom is won in realizing the true nature of self. Matter is transcended. The nature of being and the force of absolute knowledge are then revealed. ~ Patañjali

Buddha placed the mind as the central force in shaping our lives.

All that we are is the result of what we have thought. The mind is everything. What we think we become. ~ Buddha

Buddha also stated that the existence itself is ultimately illusory.

All compounded things are subject to vanish. ~ Buddha*

To someone steeped in belief of duality or materialism, Lao Tzu, Patañjali, and Buddha read like dreamy gibberish.

Conversely, those who granted these men credence ushered in religions: Taoism, Hinduism, and Buddhism, respectively. But in doing so they were, in their ignorance, turning instructive truths into myths and rituals.†

In the Western world, men seeking authority fabricated a supreme being – God – to have other men cower to the absolute power these hucksters said they were in touch with. This religious scam remains a major attraction, with suckers aplenty hoping for eternal salvation after their dreary existence here on Earth.

The religiously inclined, scientists included, take umbrage at the foregoing. But they look to self-validation, not honest investigation. The ignorant are always smug in their beliefs.

> If you would be a real seeker after truth, it is necessary that, at least once in your life, you doubt, as far as possible, all things. ~ René Descartes

Conforming to modern physics, immaterialism accords matter as proximate phenomena: an actual world in appearance, but ultimately unreal. This is admittedly difficult to believe, but it is confirmed by all that science has revealed which is otherwise inscrutable. And there is much mystery to be had, as we have already seen in our brief survey of physics, and pondering how evolutionary adaptation could possibly work other than through an intentional interactivity (teleology).

Every story is of conflict, resolved by the protagonist either gaining insight, and thereby reaching a higher plane, or else struggling onward in continuing ignorance. So, it is

---

* This quote about immaterialism suffers in translation; but its very gist remains enigmatic to those who believe that the physical world is the realm of indisputable reality.

† The usage of *ignorance* herein is of having the wrong worldview (perspective-ignorance), not of lacking facts (fact-ignorance).

that the purpose of life is to gain understanding. If it were easy, there would be no entertainment.

> We live in a fantasy world, a world of illusion. The great task in life is to find reality. ~ Irish author and philosopher Iris Murdoch

In ignorance, nattermind is given free rein; binding particulate consciousness to the mind-body. The mind creates the duplicity of duality: that one is an individual self in a physical world.

This is a deceptive separation. What you think to be the world is merely a construct of your mind.

> All that we see or seem is but a dream within a dream. ~ American writer Edgar Allen Poe

Whatever correspondence your sense of actuality has with others is possible because it is a shared belvedere – a beautiful ruse by Consciousness of physicality; the provision of the entertainment platform called *existence*.

The marvel is in how utterly convincing the illusion is. If relieving oneself of ignorance was easy, the game afoot would hardly be worth playing.

This ongoing process of incomprehension is abetted by the flames of *emotion*. Love, hate, anger, greed, and attachment exist only as appearances, like mirages in a desert of shimmering heat. Yet they are compellingly real in the moment.

To bathe in emotion is to dive headlong into the great pool of illusion. Whatever we believe is make-believe.

Nattermind's roil blocks awareness of what is, substituting for it a sense of mortal self. Removing the rubble clears the channel to unified Consciousness.

> To the mind that is still, the whole universe surrenders. ~ Lao Tzu

Whereas an unbridled mind is the source of mental illness, liberation from nattermind enables psychological health. This is *enlightenment* – the state of awake awareness with sustained quietude. The road to enlightenment is a discipline: consigning the mind to a mere utility, by exiling nattermind to the subconscious, where it may toil away.

The discipline involved in reaching enlightenment is meditation; not just as a regular practice, but as a lifestyle.

> Do not just meditate; live in meditation. ~ Nisargadatta Maharaj

Quietude emerges with persistent insistence not to indulge nattermind. In its wake comes contentment – the bliss that Nature experiences in exercising itself.

> To discover joy is to return to a state of oneness with the universe. ~ American author Pegi Joy Jenkins

The task at hand is to joyfully live. To do so carefully requires abandoning all cares, for those are the blockades to awareness of what is, built from fictive fear of what may be. Beyond the necessities of living, nary a thought is worth having.

> When the mind is free of desire and fear the unconscious becomes accessible. ~ Nisargadatta Maharaj

With the rumbling of nattermind at bay, connection to Consciousness strengthens, allowing one to comprehend Nature as it is, not merely as it appears. Enlightenment leads to *realization*: a conscious state of unity with the universe while still bodily confined.

> Try to realise it's all within yourself. ~ English musician George Harrison, in the song "Within You Without You" (1967)

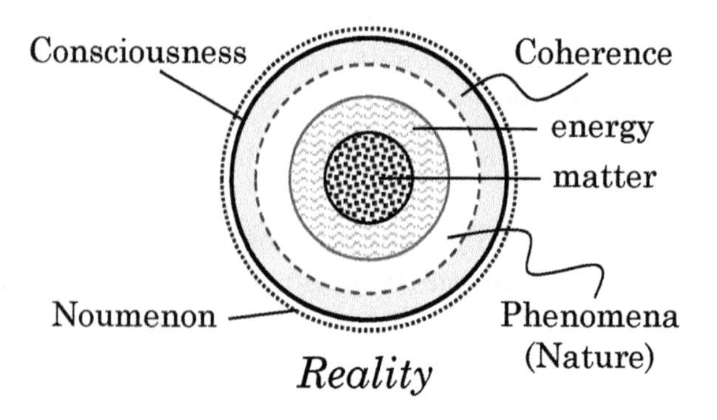

Reality

Conclusion

> Science is wonderful at destroying metaphysical answers, but incapable of providing substitute ones. Science takes away foundations without providing a replacement. ~ American philosopher Hilary Putnam

The value of science to our understanding of the world cannot be overstated. Its skepticism and method have proven invaluable tools to winnow religious nonsense from theories based upon fact.

But empirical science only goes so far. To mistake tangible actuality for the finality of reality is to fall into the arms of beliefs no different than the religious tenets which science has exposed as bogus.

All that science has revealed shows that materiality is not the foundation of reality. Quanta being coherent energy fields indelibly indicate that the seeming solidity of existence is ultimately illusory. The nature of the atom corroborates this, and the power of the mind-over-matter confirms it. Moreover, matter and energy alone cannot explain the order that pervades phenomena.

> Some secrets of Nature are hidden, although they are really close to us. ~ German taxonomist Peter Jäger

We naturally construe our lives to be "real." But what happens during a dream seems as real as events when awake. Both are equally engaging. The only difference is a sustained sense of stability, which is deceptive. The one constant of life is change. Continuity is a mirage of memory.

> The architecture of the universe is consistent with the hypothesis that *mind* plays an essential role in its functioning. ~ Freeman Dyson

The authenticity of an ultimately immaterial unicity seems mystical. But if duality and materiality are chimerical, reasoning yanks us to the only irrefutable conclusion, however fanciful it appears.

> Once you eliminate the impossible, whatever remains, no matter how improbable, must be the truth. ~ Arthur Conan Doyle

Nature sports an unfathomable diversity. Yet it all seems to have started with a singularity of energy. Materialism cannot account for the transformation.

Matter is a transmutation of energy. Energy is nothing but a constructive abstraction. Therefore, reality must ultimately be *immaterial.*

> Consciousness is a fundamental property. ~ American neuroscientist Christof Koch

The duality of existence – individual organisms within Nature – is a physical platform brought forth by a unified field of *Consciousness*. An eternal, immaterial reality sustains a material existence. Via a primordial desire for entertainment, noumenon begets phenomena.

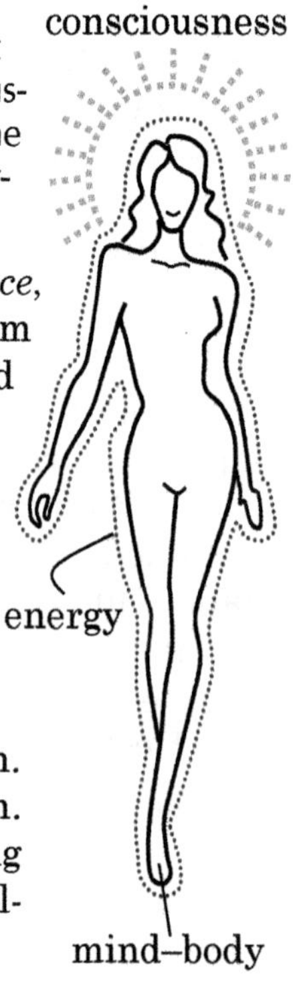

> It is not that the world does not exist. Exist it does, but merely as an appearance in Consciousness – the totality of the known manifested, in the infinity of the unknown, unmanifested. ~ Nisargadatta Maharaj

Through the natural force of *coherence*, Consciousness animates life in quantum form – seemingly individual, but always entangled within the universal field.

> The life force and the consciousness are not really two; as a concept they are treated as such, but they are really one. As soon as a form is created, the life force is infused in that form and sentience is automatically present. ~ Nisargadatta Maharaj

Matter animated by a soul is a life form. Matter not animated by a soul is a dead form. But really (as opposed to actually), nothing lives and nothing dies. If materiality is an illusion, so too is life and death.

Glossary

> The difference between the right word and the almost-right word is like the difference between lightning and the lightning bug. ~ American author Mark Twain

~: approximately.

4D (aka *spacetime*): the four dimensions of everyday experience: 3 of space (3D) + 1 of time. See *HD* and *ED*.

A

absurdism (philosophy): the inherent conflict between the human wont to find meaning and value, and the inability to truly do so.

aesthetics (aka *esthetics*): the branch of philosophy concerned with beauty.

aether (aka *ether*, *quintessence*): a long-presumed ethereal substance that pervades empty space. The assumption was eventually abandoned by physicists in the early 20th century, after a futile search. See *dark matter*.

affect (psychology): emotion.

affect heuristic: decision-making via affect.

Age of Enlightenment (aka *Age of Reason*): a cultural and intellectual movement in Western Europe during the last half of the 17th century through the 18th century that emphasized reason and individualism. The Age of Enlightenment was sparked in the late 1600s by various philosophers, including Spinoza, Locke, and Voltaire, along with Newton. Their musings promoting rationality met a receptive audience in some European rulers, who strove for *enlightened absolutism*. For one, Reason lessened the grip of the church on the body politic. The *Scientific Revolution* transpired contemporaneously with Enlightenment.

The reaction to the Age of Reason and Industrial Revolution was *Romanticism*, which originated in Europe at the end of the 18th century, and was the strong intellectual current until 1850. With an emphasis on the emotions behind the aesthetic experiences of life, Romanticism revolted against the aristocratic norms of the day, as well as taking a swipe at the scientific rationalization of the natural world, in favor of admiration and awe of the beauty and power of nature. Romanticism weaved a complex set of effects. Politically, it fostered nationalism.

alchemy: the study of matter transmutation.

algae (singular: *alga*): a eukaryotic protist, usually unicellular or colonial, that photosynthesize via chloroplasts.

algorithm: a step-by-step procedure, often employed for mathematical problems. Compare *heuristic.*

Allāh: the Islamic God.

alpha particle: two protons and two neutrons bound together into a particle identical to a helium nucleus.

altricial: species that are relatively immature and immobile at birth or hatching, and so require parental care. Many mammals are altricial. Contrast *precocial.*

Alzheimer's disease: an incurable degenerative disease leading to dementia. Symptoms advance to confusion, irritability, mood swings, trouble with language, and memory loss.

Amazon molly (*Poecilia formosa*): a small (averaging 5.5 cm) freshwater fish native to Mexico and southeast Texas, which reproduces via gynogenesis.

amino acid: an organic molecule comprising a carboxylic acid group, an amine group, and a side chain specific to the amino acid. The key elements in amino acids are carbon, hydrogen, oxygen, and nitrogen, with other elements found in the side chain.

amoeba (plural: *amoebas* or *amoebae*): a genus of protozoa comprising unicellular organisms without definite shape.

animal: a classification (kingdom) of eukaryotic heterotrophs. Most animals are motile. Other kingdoms include: archaea, bacteria, fungi, plants, and protists.

animism: the doctrine that that there is no separation between the physical and spiritual world, and that a vital force is inherent in all matter. Contrast *vitalism.*

antelope: an even-toed ungulate native to Africa and Eurasia.

anthropology: the study of human cultures and societies.

antimatter: the nearly symmetrical partner (antiparticle) to every subatomic particle.

antiparticle: an antimatter subatomic particle. The positron is the antimatter equivalent of an electron.

ape (aka *great ape*): a tailless primate; not a monkey.

arborescent (plant): a plant with wood; that is, a tree-like plant. Contrast *herbaceous* plants.

archaea: the prokaryote from which eukaryotes arose. Archaea are an extremely robust and versatile life form, with both extremophiles and ubiquity in their favor. Archaea account for 20% of Earth's biomass. As part of plankton, archaea are plentiful marine life. Social to a fault, archaea are commonly mutualists or commensals. No archaeal pathogens or parasites are known.

archipelago: an island cluster/group/chain.

arthropod: an invertebrate with an exoskeleton, a segmented body, and jointed appendages. Arachnids, crustaceans and insects are arthropods. There are presently at least 6 million different arthropods.

astrocyte (aka *astroglial cell*): a star-shaped glial cell in the brain and spinal cord.

Atlantic silverside (aka *spearing* in the northeast US; *Menidia menidia*): a small (15 cm) fish on the eastern seaboard of North America.

atom: the smallest particle of an element, comprising at the simplest a proton and an electron (hydrogen).

ATP (*adenosine triphosphate*): the cellular metabolic energy storage and intracellular energy transfer molecule. ATP is the universal cellular energy source.

Australasia: a region of Oceania comprising New Guinea, Australia, New Zealand and neighboring islands.

autophagy (aka *autophagocytosis, macroautophagy*): the breakdown and recycling of cellular components for cleanliness and nutritional reasons.

autopoiesis: a system capable of maintaining and reproducing itself. The term was coined by Chilean biologists Humberto

Maturana and Francisco Varela in 1972 to define the self-maintaining chemistry of living cells.

autotroph: an organism that produces more biomass energy than it uses during metabolic respiration. Photosynthetic life, such as plants, are autotrophs. Contrast *heterotroph*.

avian: relating to birds.

axiom: a self-evident truth requiring no proof.

awareness: the quality of being conscious in the present moment.

B

bacteria (singular: *bacterium*): a classification of prokaryotes, abundant in most ecosystems. Bacteria play vital roles in various facets of the biosphere.

bark (plants): the outermost layer of a woody plant. Bark is a nontechnical term for the various tissues outside the vascular cambium. On older stems, *inner bark* is living tissue, while *outer bark* is dead tissue.

bark beetle: a beetle of 220 genera and 6,000 species that reproduces in the inner bark of trees.

baryon: a composite particle of ordinary matter – protons and neutrons, which each consist of three quarks. See *light matter*. Contrast *dark matter*.

BCE: acronym for "Before the Common Era"; an alternative designation for the calendar era introduced by 6th-century Christian monk Dionysius Exiguus, who respectively used BC (before Christ) and AD (anno Domini) to indicate times before and after the life of Jesus of Nazareth. Year zero is unused in both systems. Dates before 1 CE are indicated as BCE. CE (common era) dates are typically not denoted, with implicit understanding that such dates are of relatively recent times.

bear: a large carnivorous mammal widespread throughout the world, mostly in the northern hemisphere. There are only 8 extant species of bears.

beauty: qualities which excite pleasure.

beetle: an order of insect (*Coleoptera*) with wings and shell-like body protection.

belief: confidence in an abstraction as truth.

Bell's theorem: a theorem by John Stewart Bell that quantum mechanics must necessarily violate either the principle of *locality* or *counterfactual definiteness*. Bell held that locality is violated and counterfactual definiteness applies.

Bible, The: a collection of ancient texts held sacred in Judaism and Christianity.

Big Bang: the theory that the universe began with an initial energetic cosmic explosion from a singularity.

biofilm: a colony of prokaryotes encased in a stabilizing polymer matrix; commonly known as *slime*.

biology: the science of life.

biota: the flora and fauna of a region.

bird: a class of feathered, bipedal, endothermic, egg-laying vertebrates. Birds descended from maniraptoran theropod dinosaurs. 10,000 living species are known.

black body: an idealized opaque/non-reflective object which absorbs all incident electromagnetic radiation. The term was coined by Gustav Kirchhoff in 1862.

black-body radiation: an electromagnetic radiation about a *black body*. Black body radiation has a specific spectrum and intensity that depends only on the temperature of the body.

black hole: a star so dense that it draws in light, rendering it black.

bliss: a natural feeling of joy in being alive.

bonobo (*Pan paniscus*): a peaceable ape, closely related to the chimpanzee and human species. Bonobos have a matriarchal society. Bonobos are notably fond of sexual behaviors.

boson: the quantum particles that carry fundamental forces under the *Standard Model*; named after Satyendra Bose. Contrast *fermion*.

botany: the study of plants.

bower: an attractive architectural display.

bowerbird: a medium-sized passerine, of 20 species, found in the Pacific region. Male bowerbirds construct elaborate bowers to attract and seduce mates.

braneworld: an HD model using branes. Braneworld models are extensions from earlier *M-theory* and *D-brane* model constructs.

Bronze Age (roughly 3300 – 1300 BCE): the middle principal period of the *three-age system*, noted for the metallurgical production of bronze. The *Stone Age* preceded, the *Iron Age* followed.

Buddhism: an offshoot religion of Hinduism, founded upon the teachings of Buddha.

BYA: billions of years ago. BY as an acronym for "billion years" is deprecated in modern geophysics, in favor of *Ga*, shorthand for *gigaannum*; go figure.

C

C_4 plant: a plant that produces oxaloacetic acid, with 4 carbon atoms, as its first-stage photosynthetic product.

cactus: a spiny plant of 127 genera and over 1750 species.

Cameroon: a country on the mid-west coast of Africa.

capitalism: an economic system based upon private ownership of resources and their exploitation for exclusive profit.

Carboniferous (359 – 299 MYA): a period during the Palaeozoic era, following the Devonian period and preceding the Permian. Vast forests covered the land. Their demise produced the coal beds which characterized the geology of the period, and after which the period is named. Amphibians were dominant. Arthropods were ubiquitous.

catalyst: a substance (molecule) that causes a change in rate of a chemical reaction by lowering the energy necessary to effect a reaction.

categorize (aka *classify*): to arrange or organize via criteria.

CE: see BCE.

cell (biology): the basic physical unit of living organisms.

Chasmagnathus granulatus: a small (0.2–3.7 cm), intertidal burrowing crab that despises seagulls.

chemistry: the study of matter, especially chemical reactions.

chimpanzee (*Pan troglodytes*): a medium-sized ape, closely related to bonobos and humans.

China: the largest country in east Asia, with the world's greatest population: 1.4 billion people in 2016. Over 90% of China's people live in the eastern half of the country, which has most of the major cities and nearly all arable land. China has one of the oldest extant civilizations.

chlorophyll: the green biomolecule in cyanobacteria and the chloroplasts of algae and plants that absorbs light for photosynthesis.

chloroplast: the photosynthetic organelle (plastid) found in algae and plant cells.

Christianity: a religion based upon hero worship of Jesus of Nazareth as the supposed son of God.

chromosome: an elaborately coiled package of genetic material within a eukaryotic cell.

cilium (plural: *cilia*): a hair-like protuberance from a cell; employed for sensory perception and/or locomotion (motile cilia). Flagella and motile cilia comprise a group of organelles termed *undulipodia*. Compare *flagellum*.

clade: a group of biological taxa that includes all descendants of a common ancestor.

cladism (evolutionary biology): categorization based upon shared characteristics.

cognition: the process of understanding, involving both awareness and judgment.

Collective: people who follow their biological urges as natural imperative, and value their emotions. As believers in taking existence at face value, the Collective are naïve empiricists.

compound: a combination of elements bonded into a molecule.

concept (aka *idea*): an abstract mental construct involving discriminatory categorization.

conceptualization: the mental process of resolving one or more perceptions into a concept (symbolic abstraction).

confirmation bias: the tendency to search for, interpret, and prioritize information in a way that confirms a held hypothesis or belief.

conscious: thoughts and desires of which one is aware. Compare *subconscious*.

consciousness: the platform for awareness in an individual life constituent (e.g., organism, cell). Compare *Consciousness*.

Consciousness: the unified field of awareness responsible for existence. Compare *consciousness*.

continent: a gigantic landmass, 7 of which are currently extant on Earth: Africa, Antarctica, Asia, Australia Europe, North America, and South America.

continued influence effect: the tendency to believe previously learned misinformation even after learning of its falsity. The continued influence effect is one aspect of *confirmation bias*.

convergent evolution (aka *parallel evolution*): the independent evolution of one or more similar traits in organisms of different clades.

coral: a colonial marine invertebrate comprising numerous identical polyps.

cosmic inflation: a myth about the early cosmos, claiming that the universe had a near-instantaneous massive inflation $3x10^{-36}$ seconds after the onset of the Big Bang. Cosmic inflation outrageously violates basic physics.

cosmogony: a conjecture as to the origin of the universe.

counterfactual (physics): values which could have been measured, but were not. This is a different definition than normal, where *counterfactual* means contrary to the facts.

crab: a ten-footed (decapoda) crustacean, typically with a thick exoskeleton and a pair of claws on its front legs.

crustacean: a large group of arthropods, including barnacles, krill, crabs, crayfish, shrimp, and lobster. There are at least 67,000 species, from 0.1 mm to 3.8 meters in size. Most crustaceans are aquatic, but some, such as woodlice, are terrestrial.

Cryogenian (850 – 635 MYA): the middle period of the Neoproterozoic era, following the Tonian and preceding the Ediacaran. A period of global glaciation (*Snowball Earth*), to which the name refers.

crystal: a solid characterized by an orderly, repeating 3d pattern. A *lattice* is a typical crystalline pattern.

curse of knowledge: a cognitive bias of assuming that others know what one knows. The curse of knowledge is revealed when someone, such as a teacher, presents inscrutable information, mistakenly presuming that the listener has the background needed to understand what is being presented.

cyanobacteria: photosynthetic eubacteria; often called *blue-green algae*, though they are not in the same group as algae.

cyclic cosmology: a model that posits the universe as eternal. The cyclic model supposes a *multiverse*.

D

D-brane: a higher dimensional (HD) object; related to *M-theory*.

Dark Ages: the 5th–10th centuries in Europe; the early Middle Ages following the decline of the Roman Empire. Coined by Francesco Petrarch in the 1330s, when writing of the previous historical period:

> Amidst the errors there shone forth men of genius; no less keen were their eyes, although they were surrounded by darkness and dense gloom. ~ Francesco Petrarch

The term *Dark Ages* is generally disparaged by contemporary historians, owing to its critical overtone. Yet the aptness of its cultural attribution cannot be denied.

dark matter: a hypothesis of matter that exists only ED, lending only gravitational distortion to 3D space. Despite relentless search for the source of its indisputable effect, no evidence of dark matter has been found. Contrast *baryon, light matter*.

desire: mental want. See *motivation*.

determinism: belief in cause and effect, from which comes the doctrine that all facts and events exemplify natural laws.

diffraction: wave fronts that modulate when passing on the edge of an opaque object, causing a redistribution of energy within the front.

diploid: an organism having two sets of chromosomes. Most eukaryotes are diploid: two sets, one from each parent, typically twined through sexual reproduction. Humans are diploid.

Dirac equation: a relativistic quantum mechanical wave equation that characterizes the spin of fermions. Created by Paul Dirac in 1928.

DNA (*deoxyribonucleic acid* ($C_5H_{10}O_4$; $H\text{–}(C{=}O)\text{–}(CH_2)\text{–}(CHOH)_3\text{–}H$)): a double-stranded molecular chain that acts as a template to build cellular components.

dopamine ($C_8H_{11}NO_2$): a hormone and neurotransmitter; associated in mammals with reward-motivated behavior.

Doppler shift (aka *Doppler effect*): a change in observed frequency relative to the source of a generated wave; proposed by Christian Doppler in 1842.

dualism: the metaphysical belief that existence is bifurcated between the physical and the mental (or spiritual). Contrast *monism*.

E

$E = mc^2$: an equivalence of energy and mass, embodying the concept that the mass of an object is a measure of its energy content. Formulated by Albert Einstein in 1905.

Earth: the third planet from the Sun; the densest and fifth-largest.

ecology: a subdiscipline of biology, concerned with the pattern of relations between organisms and their environment.

economics: the study of production, distribution, and consumption of goods and services, and of the material wellbeing of humans.

ecosystem: the community of organisms (*biota*) in a biome, and the abiotic (non-living) elements within the area.

ectotherm: an animal species without internal means to maintain thermal homeostasis. Ecothermic species, such as reptiles, practice behaviors to regulate body temperature, like lying in the sun to warm themselves. Commonly misnamed *cold-blooded*, ectotherms' blood is just as warm as endotherms. Compare *endotherm*.

ED: extra dimensions (or extra-dimensionality). ED refers to the extra spatial dimensions beyond the 3 of space (3D) that are perceptible and measurable. See *4D* and *HD*.

Egyptian (civilization) (3150 – 30 BCE): an ancient civilization in Northeastern Africa, concentrated along the lower Nile.

electromagnetic spectrum: a continuum of increasing energy intensity, from longer wavelengths to shorter.

electromagnetism: one of the fundamental physics forces, affecting particles that are electrically charged.

electron: a negatively charged subatomic particle (fermion). An electron has 1/1836 the mass of a proton when at rest. But an electron is never at rest, so the figure is just a mathematical extrapolation.

embryo: an early stage of development in multicellular diploid eukaryotes (e.g., plants and animals that sexually reproduce).

emotion: a feeling evolved into a sustained mental state.

empirical: derived from experience.

empiricism (epistemology): the presumption that knowledge derives solely from sensory experience.

empiricism (philosophy of science): the belief that the natural world may be entirely explained by physical forces.

endosymbiont: an organism living within another organism, forming a mutually advantageous arrangement.

endotherm: an animal with internal means to maintain thermal homeostasis. Birds and mammals are endotherms. Endothermy raises an animal's metabolic needs compared to ectothermic animals. Compare *ectotherm*.

energy (physics): what it takes to put matter to work; the foundational construct of existence, including the producer of matter. Energy is itself ethereal.

enlightenment (aka *quietude* or *quiet consciousness*): the state of consciousness with abiding connection to the unicity of nature. In enlightenment there is inherent contentment, accompanied by an eminently sensible perspective on life (and death). Compare *realization*.

entanglement: spatially distinct matter acting synchronously. Entanglement defies *locality*.

entropy (physics, particularly thermodynamics): the tendency of energy to dissipate and equilibrate. A measure of thermal energy unavailable for work. Introduced by Rudolf Clausius in 1850. An entropic interaction is one where energy is locally lost. Gravity is entropic.

enzyme: a protein that facilitates the activities of other proteins or substrates. Enzymes typically act as catalysts.

epidemiology: the study of health and disease in populations.

epidermis: the outermost layer of cells (in animals, the skin).

epigenetics: a heredity mechanism via gene regulation, without changing the structure of the gene involved (that is, without genetic mutation).

epiphyte: a plant that grows harmlessly on another plant, typically a tree. Epiphytes grow on other plants for physical support.

epistemology: the study of knowledge, particularly its origin, nature, methods, and limits.

ethics (aka *moral philosophy*): a branch of philosophy concerning systemizing distinction between right and wrong action; a system of moral principles.

eukaryote: an organism with cell structures separated by membranes (*organelles*). Multicellular life is eukaryotic. Compare *prokaryote*.

event horizon: a boundary in spacetime beyond which events cannot affect an outside observer. An event horizon is typically portrayed as the "point of no return" into a *black hole*.

evolution: a distinctive change across successive generations of a population.

evolutionary biology: a subfield of biology concerned with the organic processes of evolution.

evolutionary fitness: a measure of success in populations of organisms staying alive across generations.

*extra dimensions (*ED*)*: the dimensions beyond the four (4D) that are experienced (time & 3D space). See *holistic dimensions* (HD).

F

fact: recall of an observed or experienced event. Compare *real.*

faith: belief not based on empirical evidence.

false-consensus effect: a popular social phenomenon, where people believe that their own opinions, attitudes, and beliefs are more common than they actually are.

feeling: an emotive perception. Compare *emotion.*

fermentation: a metabolic process by microbes and oxygen-starved muscle cells of converting sugar to alcohol, acids, or gases.

fermion: the elemental particles that constitute matter under the *Standard Model*; named after Enrico Fermi. Contrast *boson.*

fern: a highly successful pteridophyte that arose 360 MYA.

field: an energy associated with a spacetime point.

finalism: the belief that all events are determined by their goal.

first law of thermodynamics: the notion that the total energy in an isolated system is immutable – that energy can be neither created nor destroyed in a closed system.

fish: gill-bearing aquatic animals lacking limbs with digits. 32,000 species of fish are known. Most fish are endothermic.

flagellum (plural: *flagella*): a whip-like appendage protruding from a cell; employed for locomotion and sensory perception. Compare *cilium.*

flavor (quantum mechanics): generic term for the qualities that distinguish the various quarks and leptons.

fluid: a substance that deforms (flows) under an applied shear stress. Gases, plasmas and liquids are fluids. Contrast *solid.*

fly: a small flying insect with a single pair of wings.

fomentation: instigation of riotous activity.

force (physics) (aka *interaction*): an influence that causes a change in an object.

fractal: a set of scale-invariant self-similar iterative patterns.

framing (psychology): perceiving a situation within a certain context or from a specific perspective.

framing effect: bias from the context in which a situation is considered, typically involving personal gain or loss.

fruit fly: a fly that primarily feeds on unripe or ripe fruit. Sometimes called a "true" fruit fly, as contrasted to vinegar flies that are also called "fruit flies." Compare *vinegar fly*.

fungus (plural: *fungi*): a classification of eukaryotes that includes microorganisms such as yeast and molds, as well as macroscopic mushrooms.

G

galactic web: the interconnection of galaxies via energetic filaments.

galaxy: a gravitationally-bound (by a massive black hole) cluster of star systems and stellar remnants, swirling in an interstellar mixture of gas, dust, and dark matter.

gallfly (aka *gall wasp*): a family of ~1,300 species of wasps, named after the galls they induce on plants for larval development.

gene: a conceptual interpretation of the instructions stored in DNA or RNA polynucleotides for producing a particular biosynthetic product, typically a protein.

general relativity: a generalization of special relativity. General relativity is a geometric physical theory that treats gravity as a property of 4D spacetime, based upon the mass of objects. Under general relativity, gravity distorts 4D spacetime.

genetics: the study of genes, heredity, and variation in life forms.

genome: the complete set of genes in an organism; a storage system of instructions for construction of proteins and their cognates.

genus (plural: *genera*): a category of organisms, more generic than *species*.

ghost field: a field that affects the mass of a boson via interrelations with other bosons and fermions. Ghost fields are necessary to maintain mathematical consistency in the Standard Model. Ghost fields are conventionally construed solely as a mathematical device, and nonexistent, despite being the origin of *virtual particles*, which are presumed to exist. Such inconsistency highlights that the Standard Model is a gross approximation at best, and not genuine.

Gitterwelt: a lattice world imagined by Werner Heisenberg in 1930. Gitterwelt exists in specific crystalline structures.

glia: a catch-all cell type in the physical intelligence system of an animal. Historically construed as cells that solely supported neurons.

gluon: the boson that grips quarks together, making hadrons. The gluon is the strong force carrier.

God: the myth of an immortal supreme being, omniscient, and usually omnipotent. The concept of God is object orientation run amok; one of many delusions construed by believing in what is conceived, as contrasted to actuality, and what reasonably may be inferred via science or natural philosophy.

Gödel's incompleteness theorems: two mathematical logic theorems about the inherent limits of any mathematical system; published by Kurt Gödel in 1931. The first theorem states that all truths about the arithmetic of natural numbers cannot be proven. The second theorem, extending from the first, shows that a mathematical system cannot demonstrate its own consistency.

In 1921, David Hilbert proposed a solution to a known crisis in mathematics – early attempts to formalize the foundations of math had been found to have inconsistencies and paradoxes. Hilbert's proposal was to ground all existing theories to a finite, complete set of axioms, and then prove that these axioms are consistent. *Hilbert's program*, as it came to be known, went swimmingly well until Gödel drowned it with his incompleteness theorems.

goldenrod (*Solidago*): a genus of 100–200 species of flowering plants. Most are herbaceous, and found in North America.

goose (plural: *geese*): a large waterfowl. Some other birds have "goose" as part of their names. Distantly related birds include the generally larger swans and smaller ducks.

graviton: the hypothetical bosonic particle of gravity.

gravity: a spacetime distortion caused by mass; generally considered one of the four fundamental forces, even though the other three interactions – strong, weak, and electromagnetism – are significant to subatomic particles, whereas gravity is not.

guppy (*Poecilia reticulate*; aka *million fish, rainbow fish*): a freshwater tropical fish native to northeast South America.

gynogenesis: a form of asexual reproduction related to parthenogenesis, but with the requirement that an egg be stimulated by presence of sperm – without incorporating the sperm's genetic material – in order to develop.

gyre: a conceptual framework treating a physical system as a dynamic vortex. A gyre is characterized by its structure, qualities, thermodynamics, and interactions.

H

hartebeest (aka *kongoni*): an African antelope.

HD (*holistic dimensionality*): the totality of cosmic dimensions. HD refers to the universe having more than four dimensions (4D) (3 spatial and 1 time vector). HD = 4D + ED, where ED = extra (spatial) dimensions.

Heisenberg's uncertainty principle: see *uncertainty principle*.

heliocentrism: the theory that the Sun is the center of the solar system, around which planets orbit.

herbaceous: a plant that has leaves and stems which die down to the ground at the end of the growing season. Herbaceous plants may be *annuals*, *biennials*, or *perennials*. Contrast *arborescent*.

heterotroph: an organism that uses organic carbon for growth, but cannot fix it itself. All animals are heterotrophs. Contrast *autotroph*.

heuristic (psychology): a simple, efficient rule employed to form judgments, solve problems, or make decisions. Compare *algorithm*. See *affect heuristic*.

Higgs field: the field that imparts mass to all particles.

Hinduism: the dominant religion of India. Hinduism is based upon a compilation of diverse texts, the earliest of which date to the 7th century BCE, though most are later (late BCE). Such diversity means that Hinduism is an umbrella term, housing numerous religious offshoots.

Hiroshima (Japan): a city on the southern part of Honshu, Japan's largest island. Hiroshima is best known worldwide as the target for the first nuclear holocaust, when the United States dropped an atomic bomb on it on 6 August 1945.

homogeneous: the same at all locations. Compare *isotropic*.

hominin: the hypothesized clade that descended into humans.

horizontal gene transfer (*HGT*): sharing of genetic material between organisms. In contrast, *vertical gene exchange* is gene transfer from parent to offspring.

hormone: an organic compound intended for long-distance intercellular communication; from the Greek word for *impetus*.

hornwort: a non-vascular plant of 100–150 species, found worldwide in damp or humid locales.

Hubble's law: a cosmological observation that deep space objects are observed via a Doppler shift relative to Earth, owing to their receding (moving away) from Earth.

human: a bipedal, largely furless mammal in the *Homo* genus.

hyena: a dog-like family of carnivorous mammals, endemic to Africa.

I

ichthyology: the study of fishes.

idealism (aka *subjective idealism, empirical idealism*): the monistic immaterial epistemology that what can be known of reality is a mental construction. Compare *neutral monism.*

ignorance: a state of unknowing. There are two types of ignorance: *fact-ignorance* and *perspective-ignorance. Fact-ignorance* (*fignorance*) is not knowing salient facts related to a topic. *Perspective-ignorance* (*pignorance*) is an unenlightened state of awareness – that is, lacking knowledge about the nature of reality. Reference to *ignorance* herein is of pignorance.

illusion: something that deceives by producing a false or misleading impression of reality.

illusion of knowledge: people thinking that they know more than they do.

immaterialism (philosophy): the monistic doctrine that material objects are figments of the mind, without independent existence. Contrast *materialism.*

Impatiens frithii: a small, inconspicuous epiphyte when not displaying its bright red flowers; endemic to Cameroon.

inductivism: the traditional scientific method, attributed to Francis Bacon, of incrementally (in terms of scale) proposing natural laws to generalize observed patterns. Disconfirmed laws are discarded.

In 1740, David Hume noted limitations in using experience to infer causality. First, the illogic of enumerative induction: unrestricted generalization from particular instances to all such events. Second, the presumptiveness of conclusively stating a universal law, since observation is only of a sequence of perceived events, not cause and effect. Nonetheless, Hume accepted the empirical sciences as inevitably inductive.

Alarmed by Hume, Immanuel Kant posited *rationalism* as favored by Descartes and by Spinoza. Kant noted that the mind serves to bridge the human experience with the actual world, with its innate workings creating space, time and substance. With this, Kant trashed *scientific realism* by limiting science to tracing appearances (*phenomena*), not unveiling reality (*noumena*).

inertial reference frame: a frame of reference that perceives time and space uniformly (homogenously and isotopically), and in a time-independent manner.

influenza (aka *the flu*): an infectious disease caused by an RNA-based influenza virus.

intelligence: the ability to appropriately behave.

interaction (force): an action where two or more objects have an effect on one another. The objects are not necessarily material. They may be entirely energetic. Forces only manifest through interaction.

introspection (aka *metacognition*): (the capability of) reflectively examining at one's own thoughts and feelings.

intuition: direct apprehension. Contrast *phenomenon*.

invertebrate: an animal that is not a vertebrate.

iridescence: a play of lustrous, changing color.

Islam (religion) (aka *Muhammadanism*): the religious system founded by Muhammad and informed by the Koran, with the basic principle of absolute submission to the god Allāh.

Islam (sociology): the societies predominantly practicing Islamic religion.

isotropic: the same in all directions. Compare *homogeneous*.

J

Jupiter: the fifth planet from the Sun within the solar system; a gas giant 2.5 times the mass of all other planets in the system. Jupiter has 63 moons, one more than Saturn.

K

kelp: a large seaweed (brown algae), of which there are ~30 genera. Kelp often form dense forests which support a variety of marine animals.

kilifish: a family of small fish abundantly found in fresh or brackish waters in the Americas, and to a lesser extent in southern Europe, Africa, the Middle East, and southeast Asia.

kinematics: often referred to as *the geometry of motion*, kinematics is a branch of classical mechanics that describes the motions of bodies and systems without considering the forces that cause movement.

know: to directly perceive, and thereupon understand; to recognize the nature of.

knowledge: cognition of facts or principles about existence.

L

Laplace's demon: the hypothesis that existence would be utterly predictable to an intellect that was omniscient; posited by Pierre-Simon Laplace in 1814.

lattice: a mathematical construct of symmetrical order within a group. In physics, a lattice is a lattice-like physical model. In chemistry, a lattice is a solid arranged into a lattice.

length contraction: a moving observer perceiving the length of an object decreasing.

liana: a woody vine rooted in the soil that climbs trees to the canopy.

life: an animated form capable of ecological intelligence.

life-history variable: a trait or aspect of an organism's existence related to others; often viewed comparatively, as a trade-off with other, mutually exclusive, possibilities.

light: electromagnetic radiation visible to the human eye, at a wavelength between 380–740 nanometers.

light matter: ordinary matter. Contrast *dark matter*.

lithosphere: the outermost shell of a rocky planet. Earth's lithosphere comprises its crust and upper mantle; the portions that behave elastically over geological expanses of time.

locality: the notion that an object can only be directly influenced by its immediate surroundings. See *entanglement*.

logic: the process of chaining symbols together – from a premise to a conclusion (inference) – in a way that the linkages are agreeable to other people.

logical positivism: see *neopositivism*.

M

M-theory: a physics model that extends string theory into HD branes, postulating 11 dimensions of spacetime: 10 of space and one of time.

Macaronesia: 4 archipelagos in the North Atlantic Ocean off the coast of Europe and northwest Africa.

macrobe: non-microbial life; any organism not requiring a microscope to be seen. Contrast *microbe*.

macromolecule: a large compound molecule, commonly created by polymerization of smaller subunits into polymer chains or 3D shapes. Nucleic acids, proteins, carbohydrates, and lipids are macromolecules.

mammal: a class of air-breathing vertebrate animals, characterized by endothermy, hair, and females with functional mammary glands.

mass (classical physics): a measure of matter's inertia (resistance to a change of motion); alternately, a measure that an object of matter has gravitationally.

mass (quantum mechanics): the energy level at which an elementary particle may make an appearance in 4D.

materialism (philosophy): the monistic metaphysical belief that only matter and energy exist. Materialism supposes that the mind is a figment of... something substantial. Contrast *immaterialism*.

mathematics: the systematic treatment of relations between symbolic entities.

mechanics: the branch of physics concerned with the actions of forces on bodies and with motion.

meditation: a practice intended to achieve a state of quiet, transcendental consciousness.

megapode (aka *incubator bird*): a stocky, chickenish bird with a small head and large feet; endemic to Australasia.

mentation: mental activity.

mesophyll: the parenchyma, usually containing chlorophyll, which forms the interior parts of a leaf.

Mesopotamia: an area of the Tigris–Euphrates river system; widely considered the Western cradle of civilization during the Bronze Age. Indigenous Sumerians, Assyrians, and Babylonians were there at the onset of written history 3100 BCE.

metaphysics: philosophy concerned with first principles, including ontology and epistemology.

metabolism: cellular chemical reactions which provide energy for vital processes.

microbe: a microorganism; an organism too tiny to be seen without a microscope; often a single-celled *prokaryote*. Microbes include archaea, bacteria, and fungi. Contrast *macrobe*.

microbiome: the ecological community that comprises every eukaryotic organism, especially multicellular eukaryotes. Commensal microbial prokaryotic inhabitants are essential to eukaryotic life.

mind: an intangible organ for symbolic processing.

mind-body: the mind & body as an integral life form.

mind-body problem: the unsolvable inquiry into the functional interface between the intangible mind and the physical body.

mitochondrion: an organelle that acts as a cell's power plant, generating a supply of *ATP*.

monism: the belief that there is a singular reality, whether *materialism* or *immaterialism*. Contrast *dualism*.

morality: the differentiation between right and wrong based upon fairness. The philosophy of morality is *ethics*. A *moral code* is a creed of morality.

moss: a small, non-vascular plant that typically grows in a clump. There are ~12,000 moss species.

motivation: a stimulus that causes an organism to behave in a certain way. See *desire*.

multiverse: the idea that a multitude of universes exist HD on a vast canvas of endless time. Based upon erroneous assumptions, mistaking wonky math for reality, many multiverse models are nonsensical. See *cyclic cosmology*.

mutation: a change in a genetic sequence.

mysticism: the doctrine that knowledge of ultimate reality may be subjectively intuited. Mysticism is true, with the caveat that subjectivity is ultimately an illusion. (Can illusions be true?)

N

naïve realism: the belief that one's perception of actuality is reality, objectively and without bias.

natural philosophy: the study of Nature from a holistic perspective; the common methodology of inquiry into Nature prior to the 17th century. Contrast *science*.

natural selection: a disproven hypothesis of evolutionary descent by Charles Darwin, who proposed that random mutations over millions of years led to speciation, whereupon new species survived or went extinct by competition. Senselessly, this meaningless term remains popular among evolutionary biologists.

> Natural Selection almost inevitably causes much Extinction of the less improved forms of life. ~ Charles Darwin

Nature: the exhibition of existence.

nature (of): the essence; the basic constitution.

neonate: a newborn offspring.

neopositivism (aka logical positivism): The idea that there are no valid ideas; that only empirically verifiable observations can be considered cognitively meaningful. Influenced by the field theories of physics in the early 20th century, and most particularly under sway of Ernst Mach, neopositivism arose among Viennese intellectuals in the 1920s. Rejecting metaphysics, neopositivism's central creed is that only empirical facts form valid knowledge. See *verificationism*. Contrast *panpsychism*.

neuron (aka *nerve cell*): an electrically excitable intercellular signaling cell which is an integral part of a nervous system. See *glia*.

neutron star: a stellar remnant from the gravitational collapse of a massive star (supernova). Neutron stars are mostly made up of neutrons, condensed to the utmost extent.

neutral monism (aka *neumonism*): the immaterial epistemology that the essence of existence is neither material nor mental, but energetic. Compare *idealism.*

neutron: a subatomic particle at home in the nucleus of an atom. Lacking an electromagnetic charge, neutrons act as a peacemaker in holding feisty protons together in an atomic nucleus. See *proton.*

neutron star: a stellar remnant from the gravitational collapse of a massive star (supernova). Neutron stars are mostly made up of neutrons, condensed to the utmost extent.

nonlocality: entanglement of objects at some distance from each other. Contrast *locality.*

noumenon: nonexistence; outside of existence. In western philosophy, a *noumenon* is beyond perception, as contrasted to *phenomena.*

nucleic acid: a large biomolecule (biopolymer) essential to life; discovered by Friedrich Miescher in 1869. DNA and RNA are nucleic acids.

nucleotide: an individual structural unit of nucleic acids. A nucleotide is a nucleobase packaged with sugar and phosphate groups, held together by ester bonds.

nucleus (physics): the central core of an atom, comprising protons and neutrons.

O

olfaction (aka *oflactics*): the act or sense of smell.

olfactory bulb: a vertebrate neural bundle involved with smell.

omniscience: having complete knowledge.

ontology: the branch of metaphysics concerning the nature of existence.

organ (aka *viscus*) (biology): a collection of interconnected tissues dedicated to a common function.

organelle: a subunit within a eukaryotic cell that has a specialized function. Organelles are membrane-bound. Cell organelles evolved through endosymbiotic union with an archaeon host cell and a bacterial endosymbiont.

organic: related to living organisms. From a chemistry viewpoint: a complex molecular structure based upon a carbon backbone.

organism: a life form.

oxygen (*O*): the element with atomic number 8; a highly reactive nonmetallic element that readily forms compounds (notably oxides) with almost all other elements. Oxygen is the third most common element in the universe.

P

paleoanthropology: the study of hominins from physical evidence. Paleoanthropology combines *paleontology* and *anthropology*.

paleontology: the study of prehistoric life.

panpsychism: the philosophic view that consciousness, soul (psyche), and mind are part of all life. Panpsychism was the prevailing orthodoxy until the mid-20th century, when supplanted by *neopositivism*.

parenchyma (botany): the most common and versatile ground tissue in plants, composed of thin-walled cells able to divide.

parrot: an uncommonly intelligent bird of 86 genera and 372 species, found in many tropical and subtropical biomes. The greatest parrot diversity is in Australasia and South America.

parthenogenesis: a form of asexual reproduction, where an unfertilized egg cell nonetheless develops into an embryo. Sperm or pollen may trigger embryonic development without making a genetic contribution. See *gynogenesis*.

particle (physics): a point in spacetime, typically used to ascribe a quantum-sized localized field. Contrast *wave*.

passerine (bird): a bird in the *Passeriformes* order, comprising over half of all bird species: over 5,000 identified species in over 110 families. One of the most diverse terrestrial vertebrate orders, around twice that of the largest mammal order: rodents. Passerines include most perching birds, such as sparrows, wrens, finches, tits, and corvids.

Pauli exclusion principle: a theoretical requirement that two fermions cannot occupy the same space simultaneously. Formulated in 1925 by Wolfgang Pauli.

peptide: a short chain of amino acids – 2 to 50 or so. A longer chain is properly termed a *protein*.

perception: the process of mentally integrating sensory input (*sensation*) using memory, thereby creating a composite representation which is usually comprehensible. Compare *conceptualization*.

periodic table of elements: a tabular display of atomic species (chemical elements), presented in increasing order of their *atomic number* (number of protons), with columns (groups) and rows (periods) based upon electron configuration.

phantom limb: the sensation in animals that a missing limb is still attached. Phantom sensations are experienced in many body parts, including lost eyes and extracted teeth.

phenomenon: an observable event. Contrast *noumenon*.

pheromone: a secreted or excreted hormone employed as a communication signal.

philology: the study of language in historical texts.

philosophy: the study of fundamental abstractions, including the natures of reality, existence, knowledge, rationality, the mind, morality and values. The term *philosophy* derives from the ancient Greek for "love of wisdom." There are three branches of philosophy: natural, moral, and metaphysical. *Natural philosophy*, which evolved into *science*, concerns Nature. *Moral philosophy* deals with the principles of ethics. *Metaphysics* considers first principles, such as ontology, and is intimately connected with epistemology.

phloem: tissue that distributes sugar-laden sap among a plant. Compare *xylem*.

photoelectric effect: a body's emission of charged particles (photoelectrons or ions) upon absorbing electromagnetic radiation.

photoelectron: an electron emitted from an object via the photoelectric effect.

photon: a hypothetical bosonic quantum of light. Commonly called a *packet* of light energy, as light exists in an apparent duality of particulate and wave forms. Though photons do not interact with each other, they somehow manage to carry the force of *electromagnetism*.

photosynthesis: converting sunlight into energy by a living organism.

phototroph: an organism that can turn light energy into metabolic chemical energy.

phototropism: a natural tendency for light to be an orienting stimulus.

physical model: a mathematical model, typically geometric or algebraic, providing a mathematical description of the embodied phenomena.

physical theory: an explanation of relationships between various measurable phenomena. A physical theory may include a model of physical events (i.e., a *physical model*).

pilot wave theory: a theory that there is an inherent wave-particle duality for every elementary particle.

placebo: a simulated medical treatment intended to deceive the recipient and thereby provoke the *placebo effect* of actually working to relieve or even cure the targeted affliction. The term *placebo* originated with an old Latin word for "I shall please." In medieval times, a placebo opened the Catholic Vespers for the Dead, which were sung by hired mourners for a funeral; sycophants who wept crocodile tears on behalf of the family. This gave placebo the odious meaning of a toady. The term took its medical context in the early 19th century, when placebos were remorsefully employed. Placebos' efficacy gradually transformed their moral worth.

Planck constant (aka *Planck's constant, Plank's action quantum*): a physical constant reflecting the size of energy quanta in quantum field theory. Planck's constant states the proportionality between the momentum and quantum wavelength of every subatomic particle. The relation between the energy and frequency of quanta is termed the *Planck relation*.

Planck length: the minimal theoretical limit to spatial distance; a measure derived from Newton's gravitational constant, the speed of light in a vacuum (*c*), and Planck's constant. Planck length is 1.616199×10^{-35} meters.

Planck time: the theoretical limit of temporal measurement; the time required for light in a vacuum to travel a single Planck length. At 5.391×10^{-44} seconds, Planck time is the shortest sprint imaginable.

Planck unit: a system of natural units used in physics, particularly *Planck length* and *Planck time.*

plant: a categorization (kingdom) of autotrophs, including mosses, ferns, conifers, and flowering plants (angiosperms).

pluralistic ignorance: a social phenomenon where most members of a group do not believe what they consider to be the majority outlook or belief of the group.

plutonium (Pu): the element with atomic number 94; a silvery-gray metal that tarnishes when oxidized, much like nickel. Plutonium is the heaviest primordial element.

polarization (optics): a state of light in which the radiation exhibits different properties in different directions.

precocial: species with relatively mature and mobile young from the moment of birth or hatching. Many, though not all, arthropods, fish, amphibians, and reptiles are precocial. Contrast *altricial.*

prickly pear cactus: a cactus in the genus *Opuntia*, native to the Americas.

primary metabolite: a compound produced by a plant that is essential to its survival. Compare *secondary metabolite.*

primate: a mammal order containing prosimians (neither monkey nor ape) and simians (monkeys and apes).

prokaryote: an organism that lacks a cell nucleus or other membrane-bound organelles. While prokaryotes are strictly single-celled, most are capable of forming stable communities, such as a *biofilm*. Contrast *eukaryote.*

protozoan (*plural*: protozoa): one of a diverse group of unicellular eukaryotic organisms, many motile.

protein: a single, long, linear polymer chain of amino acids that typically takes a folded structure.

protist: a catchall kingdom of eukaryotic organisms that includes algae and amoeba.

proton: a positively charged hadron that is a constituent of every atomic nucleus. The simplest hydrogen atom comprises a proton nucleus with a single electron orbiting about it. See *neutron*.

pteridophyte: a vascular plant that reproduces and disperses via spores, producing neither flowers nor seeds.

Q

quantum (plural: *quanta*): an infinitesimal (Planck-sized) chunk of ripple in an energy field that appears as a particle.

quantum effect: a physical 4D effect reflecting HD dynamics. Entanglement is a quantum effect.

quantum field theory (*QFT*; aka *quantum theory*, *quantum mechanics*): a theoretical framework explaining subatomic interactions from a particle perspective.

quantum foam: the characterization of the ground state as a froth of virtual particles continually perturbed by ghost fields of tremendous vacuum energy.

quantum mechanics: see *quantum field theory*.

quasar: a cosmic light source caused by the spin-off of a black hole.

quasiparticle: an emergent approximation of fermionic behavior. Localized subatomic energies which mimic bosons are termed *collective excitations*.

quorum sensing: decision making in a decentralized network.

R

rare-male effect (aka *negative frequency-dependent selection*): the process in which the evolutionary fitness of a trait goes up as its relative abundance goes down.

rational (psychology): agreeable to reason, good sense, and sound judgment.

rationalism: the philosophic belief that reason is the source and arbiter of knowledge.

> All our knowledge begins with the senses, proceeds then to the understanding, and ends with reason. There is nothing higher than reason. ~ Immanuel Kant

real: that which exists (actuality), as contrasted with what may be experienced. See *truth*, *reality*.

reality: that which necessarily is (phenomenal or noumenonal), neither dependent nor derivative.

realization (aka *unity consciousness*): the state of consciousness with abiding experience of the unicity of Nature. Compare *enlightenment*.

reductionism: the hypothesis that every complex phenomenon can be explained by analyzing the most basic phenomena in play during a relevant event. Reductionism requires that the something can never be more than the sum of its parts. Reductionists explain biological processes in the same way that chemists and physicists interpret inanimate matter. Reductionism adheres to empirical cause and effect with rigor, and so is a tool supporting materialism. See *synergy*.

refraction: a change in propagation direction of a wave due to a change in its transmission medium.

relativity of simultaneity (physics): the concept that distant simultaneity – two spatially separate events occurring at the same time – is not absolute; instead, dependent upon the observer's perceptual frame of reference.

retina: the light-sensitive layer of tissue lining the inner surface of the eye.

ribosome: the cellular factory for synthesizing proteins from peptide pieces.

risk sensitivity: the capability of an organism to discriminate between stable and unstable environments.

RNA (*ribonucleic acid* ($C_5H_{10}O_5$; H–(C=O)–$(CHOH)_4$–H)): a macromolecule comprising a long chain of nucleotides. RNA & DNA differ by their sugar (ribose versus deoxyribose (a ribose lacking an oxygen atom)). RNA & DNA also differ by one nucleobase – whereas RNA uses uracil (U), DNA employs thymine (T). See *DNA*.

rodent: an order of mammals characterized by constantly growing incisors that must be kept short by gnawing. ~40% of mammal species are rodents; 2,227 known species.

S

saltation (biology): a sudden evolutionary change from one generation of organism to the next.

schizophrenia: a severe mental disorder characterized by mistaking unreality for actuality.

science: the study of Nature from a strictly empirical standpoint. William Whewell coined the term *scientist* in 1840. Contrast *natural philosophy*.

scientific realism: the naïve belief that actuality is reality.

seagull: a medium to large seabird, usually white or gray.

secondary metabolite: a specialty compound produced by a plant for ecological purposes. Compare *primary metabolite*.

self-esteem (aka *self-worth*, *self-regard*, *self-respect*): individual emotive assessment of one's own quality.

sensation: the process of receiving stimuli from sensory organs for collation and interpretation via *perception*.

Serengeti: a grassland plain ecosystem in Africa.

shear-thinning liquid: a liquid with non-Newtonian behavior, in which viscosity lessens under shear strain. Shear-thinning liquids are typically polymers, not pure liquids with low molecular mass.

sinusoidal wave: the distance between adjacent peaks or troughs of a measured energetic waveform.

sleepwalking (aka *somnambulism*, *noctambulism*): a sleep disorder of combined sleep and wakefulness, where people asleep perform activities usually done only while awake.

snapdragon (*Antirrhinum majus*): a species of flowering plant native to the Mediterranean region.

sociality: the general affinity towards others, especially of the same species.

solid: a substance with structural rigidity. Crystals and glasses are solids. Contrast *fluid*.

solipsism: the irrefutable argument that only the self can be proven to exist.

soul: the part of an organism capable of passively witnessing its own mentation (introspection); more transcendentally, the perpetual essence of an organism.

space: a boundless, non-Euclidean extent as filler for celestial bodies, which are invariably in motion.

spacetime: a treatment of space and time via unified dimensionality.

special relativity: a physical theory of measurement proposed by Albert Einstein in 1905: that the speed of light provides an inertial frame of reference. Special relativity has numerous consequences beyond uniform motion being relative, including relativity of simultaneity, time dilation, and length contraction.

speciation: the process of species formation.

species (biology): a population that does not generally interbreed with another population of similar organism. Though there may be no biological impediment to breeding, organisms may choose not to interbreed by preference, and hence are considered separate species. Ernst Mayr is generally credited with the modern definition.

> *Species* are groups of actually or potentially interbreeding natural populations, which are reproductively isolated from other such groups. ~ Ernst Mayr in 1942

species (chemistry): chemically identical molecular entities with distinct interaction characteristics, typified by different ionization or lack thereof.

spin (quantum physics): the intrinsic angular momentum of a subatomic particle. Each particle type has a definite spin. In the *Standard Model*, only the *Higgs boson* is presumed without spin. The term *spin* is hangover from a classical physics assumption that remains unproven, and is somewhat beside the point. Spin is a mathematical property.

spontaneous symmetry breaking (SSB): a mathematical concept wherein the manifestation of a symmetrical system shows a tangible result, which breaks symmetry merely by actualization. The system may remain symmetrical (hidden symmetry), but its outputs never are, as symmetry has to be broken for any realized result to be had.

"spooky action at a distance": Einstein's dismissive term for *entanglement.*

stamen: the (male) pollen-producing organ in a flower; also termed *androecium*. The stamen has a stalk (*filament*) and an *anther* that contains pollen (microsporangia). See *stigma.*

standard cosmological model (*ΛCDM* or *Lambda-CDM*): an inaccurate model of cosmogony and cosmology, adhering to the notions of the Big Bang and cosmic inflation, with cosmic expansion presently accelerating.

Standard Model: a quantum field theory focused on fundamental subatomic particles and their interactions.

star: a massive, luminous sphere of plasma held together by gravity.

statistical mechanics: a branch of theoretical physics using probability theory to study mechanical systems. Modeling irreversible processes driven by imbalances is termed *non-equilibrium statistical mechanics*. Such processes include chemical reactions, thermodynamics, and particle flows.

stigma (botany): the (female) portion of a flower that receives pollen during pollination. A pollen grain germinates on the stigma, which is often sticky. The tube-like *style* connects the stigma to the ovary. See *stamen.*

stomata (singular: *stoma*): plant pores.

stria (plural: *striae*): a narrow furrow, stripe, streak, or ridge.

striation: employment of striae in a parallel arrangement.

string theory: a theoretical attempt to reconcile quantum field theory with general relativity by characterizing quanta through their vibrational quality.

strong force: the force binding quarks and antiquarks to make hadrons, as well the nuclear force gripping protons and neutrons together in atomic nuclei. Compare *weak force.*

subconscious (aka *unconscious*): mentation that one is not aware of (not conscious of); below the threshold of the conscious awareness. Compare *conscious.*

Sun: the star at the center of the solar system, with a diameter of 1,392,000 km.

sunbird: a small, slender, Old World passerine, usually with a downward-curving bill.

sundew: a parasitic plant in the genus *Drosera*, of which there are at least 194 species. All sundews lure, capture, and digest insects via adhesive-tipped glands on stalks grown out of leaf surfaces. The ingested insects compensate for the poor mineral nutrition of the soil in which sundews live.

sunflower (*Helianthus annuus*): an annual plant native to the Americas. Sunflowers are notable for their large flowering head.

superconductivity: zero electrical resistance, resulting from electrons overcoming their mutual repulsion and pairing up, creating a coherent, frictionless flow.

supersymmetry (SUSY): a unifying field hypothesis for fermions and bosons, bringing together all quantum particles as components of a single master superfield. SUSY lacks any evidentiary basis, as requisite partner particles have not been found.

symbiosis: two dissimilar organisms in continual interaction, often in a mutually beneficial association (*mutualism*).

synergy: an interaction of elements that, when combined, produce a total effect greater than the sum of the individual contributions. See *reductionism*.

T

Taoism: a Chinese religious tradition emphasizing living in harmony with Nature and the ineffable Tao. Taoism dates at least to the 4th century BCE, and to the legendary Lao Tzu.

tardigrade (aka *water bear*): a 0.2–1.2 mm long aquatic animal.

tarsal (insects): the distal part of the leg, analogous to the foot.

tautology: repetition of an idea.

taxon (plural: *taxa*): a group of organisms. Taxa either have a formal or scientific name. Scientifically-termed taxa are governed by *nomenclature codes*: naming rules overseen by scientific organizations.

tectonics: processes related to the movement and deformation of the Earth's crust.

tectonic plate: a sizeable chunk of the lithosphere, including some of Earth's crust, capable of movement.

teleology: the doctrine that final causes exist; regarding biology, that evolution at some level involves goal orientation.

telomere: a protective region of repetitive nucleotide sequences at each end of a chromosome copy.

tensor: a geometric object describing linear relations between other geometric entities (vectors, scalars, tensors). A tensor is a geometric entity particularly entangled with other tensors. Yes, tensors are a tautology of entanglement.

tensor network: a network of tensors.

testosterone ($C_{19}H_{28}O_2$): a steroid hormone found in reptiles, birds and mammals; the primary male sex hormone.

theory: fact-based speculation about the relations between conceptual entities. See *physical theory*.

theory of mind: the cognitive ability to attribute mental states to oneself and others; particularly that other beings have minds.

thermodynamics: a branch of physics concerned with the dynamics of heat and temperature, and their relation to energy and work.

time: the idea that there is a temporal vector comprising past, present, and future.

time dilation: that concept that time itself is relative to the motion of an observer.

transpiration: normal release of water by plants.

tree of life (biology): a metaphor for categorizing organisms by evolutionary descent; used by Charles Darwin in 1872. The term has ancient philosophic roots.

truth: conformity with reality.

U

ultraviolet catastrophe: a classical physics prediction contravened by black-body radiation. Classical physics predicts that a black body in thermal equilibrium will emit energy equally across all wavelengths (which it does not), and emit more energy as radiation frequency upon it increases, to infinity (which contradicts the thermodynamic law of the conservation of energy).

uncertainty principle: the proposition that quantum phenomena are inherently probabilistic in their activity; a measurement may yield only an approximation of either a quantum particle's position or momentum, but not both simultaneously. This is an intrinsic property of Nature, not a measurement incapacity. Proposed by Werner Heisenberg in 1926, and controversial ever since.

ungulate: a group of mammals which use the tips of their toes, typically hoofed, to sustain body weight while moving. Commonly known ungulates include the horse, cattle, bison, camel, goat, pig, sheep, donkey, deer, tapir, antelope, gazelle, giraffe, camel, rhino, and hippo. Even-toed ungulates (*Artiodactyla*) bear their weight equally between the third and fourth toes. Odd-toed ungulates (*Perissodactyla*), which have an odd number of toes on their rear feet, bear weight on their third toe.

universal common ancestor: the notion that life arose from a single life form.

uranium (U): the element with atomic number 92; a silvery-white metal that is weakly radioactive because all its isotopes are unstable. The decay of uranium, thorium, and postassium-40 are a main source of heat in Earth's mantle, keeping the outer core liquid and driving mantle convection, which in turn facilitates plate tectonics.

V

vacuole: a membrane-bound organelle present in all plant and fungal cells, and some protist, bacterial and animal cells.

vacuum: the idea of empty space. Vacuum has been shown *not* to exist at the quantum level.

vacuum energy: the underlying energy of 4D empty space. Vacuum energy is the ground state from which 4D *virtual particles* arise. Vacuum energy is an HD phenomenon.

vascular: a life form with vessels to carry fluids; commonly used to identify land plants: *vascular plants* (aka *tracheophytes*).

vegetable: any plant whose fruit, seeds, or parts are used as food by humans; also used to refer to the edible portion of such a plant.

Venus fly trap (*Dionaea muscipula*): a carnivorous plant, native to the subtropical wetlands on the east coast of the US.

verificationism (aka *verification principle*): the epistemological doctrine that only verifiable facts are meaningful. See *neopositivism*.

vertebrate: an animal with a backbone and spinal column.

vinegar fly: a fly that lingers about overripe or rotting fruit, in the genus *Drosophila*. Confusingly, *Drosophila* are often called fruit flies. Compare *fruit fly*.

vision: the sense of sight through light.

virtual particle: a supposed subatomic particle that pops in and out of 4D; a 4D manifestation of vacuum energy. So-called virtual particles are actually fields that phase-shift in appearance between 4D and ED. In other words, virtual particles are HD. See *ghost field*.

virus: an obligate parasite that infects all other life forms.

vitalism (biology): the doctrine that there is a vital force related to living organisms, distinct from chemical and physical forces.

vitalism (natural philosophy): the doctrine that living organisms are fundamentally different from non-living entities. Contrast *animism*.

W

wave (physics): a mathematical characterization of a field. Contrast *particle*.

wave-particle duality: the notion that a physical object simultaneously possesses the properties of a wave and a particle.

wavelength: the spatial period of a sinusoidal wave; commonly used as a statistical measure of the energy of a waveform, which is mathematically the product of a wave's frequency and amplitude.

weak force: the nuclear force that changes one variety of matter into another; responsible, *inter alia*, for beta decay; transmitted by the W or Z boson. Compare *strong force*.

weight: the force, measured in newtons, that gravitation exerts upon an object or body, equal to the mass (m) of the body times the local acceleration of gravity (g): $W = m \times g$. In a region of constant gravitational acceleration, weight is commonly taken as a measure of mass; hence the easy confusion between *mass* and *weight*.

Weltanschauung: see *worldview*.

wildebeest (aka *gnu*): a genus (*Connochaetes*) of antelope, native to Africa, in the family of even-toed ungulates. There are two gnu: black and blue. The *blue wildebeest* remained in its original range, and so is little changed from its ancestors. *Black wildebeest* adapted to the open grassland habitat that ranges south of where blue wildebeest live.

work (physics): energy in transit; the product of an energetic force applied to matter.

worldview (*Weltanschauung*): a comprehensive conception of reality and the nature of existence.

wormhole: a shortcut in spacetime, allowing entanglement.

X

xylem: plant tissue employed to transport water up a plant. Compare *phloem*.

Z

Zeno effect (aka *Turing paradox*): a static quantum state created by continuous observation.

❧ People ☙

Adler, Mortimer J. (1902 – 2001): American philosopher and educator.

Alzheimer, Alois (1864 – 1915): German physician and psychiatrist, credited with discovering the disease that bears his name.

Anaximander of Miletus (610 – 547 BCE): Greek philosopher, astronomer, geographer, mathematician, and proponent of science.

Anderson, Philip W. (1923 –): American physicist.

Andolfatto, Peter: Canadian evolutionary biologist.

Andrulis, Erik D.: American microbiologist who works on gyre theory.

Archimedes (287 – 212 BCE): Greek mathematician, physicist, engineer, inventor, and astronomer, considered one of the leading scientists in antiquity, and one of the greatest mathematicians of all time.

Archimedes was killed during the Siege of Syracuse by an impatient Roman soldier, despite explicit orders that Archimedes was not to be harmed. The solder got ticked off because Archimedes told him to wait until he finished a problem he was working on. The soldier was executed for his indiscretion.

Arendt, Johanna "Hannah" (1906 – 1975): German-born Jewish American philosopher who philosophically rejected the label *philosopher* as being concerned with "man in the singular," preferring instead *political theorist.*

Aristotle (384 – 322 BCE): Greek philosopher and polymath.

Asch, Solomon E. (1907 – 1996): American Gestalt psychologist and social psychologist, best known for his study of conformity.

Astington, Janet Wilde: Canadian developmental psychologist.

Bacon, Francis (1561 – 1626): English philosopher, scientist, and statesman. Bacon has been called the father of empiricism.

Bahcall, Neta A.: American astrophysicist, interested in the large-scale structure of the universe.

Barrett, Jonathan: English particle physicist.

Baudrillard, Jean (1929 – 2007): French sociologist and philosopher.

Bell, John Stewart (1928 – 1990): Irish physicist who developed *Bell's theorem*, which posits nonlocality (e.g. entanglement).

Ben-Jacob, Eshel (1952 – 2015): Israeli physicist, interested in self-organization, particularly among bacteria.

Bettini, Alessandro: Italian particle physicist.

Bierbach, David: ichthyologist and ecologist.

Bohr, Niels (1885 – 1962): Danish physicist who contributed to atomic theory and quantum mechanics.

Boltzmann, Ludwig (1844 – 1906): Austrian physicist who made significant contributions to mechanics and thermodynamics. Boltzmann advocated atomic theory when it was still quite controversial.

Boorstein, Daniel J. (1914 – 2004): American historian.

Boothby, Thomas: American biologist, interested in the fundamental mechanisms of extreme stress tolerance.

Bose, Satyendra Nath (1894 – 1974): Indian mathematician and physicist who worked on electromagnetic radiation and statistical mechanics.

Boulanger, Lisa: American molecular biologist.

Boutroux, Pierre (1880 – 1922): French mathematician and historian of science, best known for his accounts of the history and philosophy of mathematics.

Braakman, Rogier: Dutch chemical physicist.

Broad, William J.: American science reporter.

Bronowski, Jacob (1908 – 1974): Polish-born British mathematician, science historian, poet, and inventor.

Buddha (563 – 483 BCE): Indian guru whose teachings were the foundation of Buddhism.

Buffett, Warren E. (1930 –): American billionaire investor.

Caetano-Anollés, Gustavo: Argentinian biologist.

Cairns-Smith, A. Graham (1931 –): English chemist and molecular biologist.

Cajal, Santiago Ramón y (1852 – 1934): Spanish neurologist.

Calle, Carlos I.: American physicist.

Camus, Albert (1913 – 1960): French philosopher, author, and journalist. Camus' writings contributed to the rise of the philosophical school known as *absurdism*.

Capra, Fritjof (1939 –): Austrian-born American physicist and systems theorist.

Cardona, Tanai: Columbian life scientist.

Carlin, George (1937 – 2008): sardonic American comedian.

Carlson, Linda E.: American physician.

Carpenter, William T.: American psychiatrist, interested in severe mental illness, especially schizophrenia.

Carr, John P.: English plant pathologist.

Carroll, Sean M. (1966 –): American theoretical cosmologist.

Carson, Rachel (1907 – 1964): American marine biologist, famous for *Silent Spring* (1962), which chronicled the environmental devastation caused by synthetic pesticides, especially DDT. American chemical companies were incensed by the book.

Chamovitz, Daniel: American botanist.

Chandrasekhar, Subramanyan (1910 – 1995): Indian astrophysicist.

Chess, Barry: American molecular biologist.

Chunharas, Chaipat: Indian neuroscientist.

Churchill, Winston (1874 – 1965): English politician (Labour); UK Prime Minister (1940 – 1945, 1951 – 1955).

Clarke, Arthur C. (1917 – 2008): English science and science fiction writer, and futurist.

Clausius, Rudolf (1822 – 1888): German mathematical physicist who formulated the 2nd law of thermodynamics, and introduced the concept of entropy, in 1850.

Cohen, Lisa J.: American psychologist.

Cook, John: Australian psychologist.

Copernicus, Nicolaus (1473 – 1543): Prussian astronomer who developed a comprehensive heliocentric cosmology, displacing the Earth from the center of the universe. Copernicus's work was published posthumously, as he worried about the scorn that his crazy idea would provoke.

Cosby, Bill (1937 –): American comedian, actor, and author.

Costa, Fabio: quantum physicist.

da Vinci, Leonardo (1452 – 1519): Italian painter, draftsman, sculptor, architect, musician, inventor, scientist, mathematician, engineer, geologist, cartographer, anatomist, botanist, and writer. Best known for a small portrait of a drab woman with a half-smile (*Mona Lisa*).

Darch, Sophie E.: English molecular biologist.

Darimont, Chris: Canadian evolutionary ecologist.

Darwin, Charles (1809 – 1882): English naturalist, famous for his disproven hypothesis of evolution by *natural selection.*

Davidson, Richard J.: American psychologist.

Davies, Paul C.W. (1946 –): English theoretical physicist, cosmologist, and astrobiologist, intent on finding extraterrestrial life. Davies generated controversy by noting that the faith of scientists is in the immutability of physical laws; a faith with roots in Christian theology. Davies called the claim that science is "free of faith": "bogus."

Davis, Tamara M.: Australian astrophysicist and ultimate Frisbee player.

Dawkins, Richard (1941 –): English evolutionary biologist.

de Broglie, Louis (1892 – 1987): French physicist who developed the pilot wave theory.

de La Rochefoucauld, François (1613 –1680): French author.

de Vries, Hugo (1848 – 1935): Dutch botanist, and one of the first geneticists. de Vries coined the term *mutation.*

Descartes, René (1596 – 1650): French rationalist philosopher and mathematician.

Diderot, Denis (1713 – 1784): French philosopher, writer, and art critic; a prominent figure in the Enlightenment.

Dirac, Paul (1902 – 1984): brilliant English theoretical physicist who contributed to the early development of quantum physics. Dirac was a precise and taciturn man. Raised Catholic, Dirac once remarked, "religion is a jumble of false assertions, with no basis in reality."

Dobzhansky, Theodosius (1900 – 1975): Ukrainian geneticist and evolutionary biologist.

Doppler, Christian (1803 – 1853): Austrian physicist who proposed the *Doppler effect* in 1842.

Doyle, Arthur Conan (1859 – 1930): Irish-Scots writer and physician, best known for the crime fiction tales of detective Sherlock Holmes.

Dyson, Freeman (1923 –): English-born American physicist, cosmologist, and mathematician.

Ecker, Ullrich K.H.: Australian psychologist.

Einstein, Albert (1879 – 1955): German theoretical physicist, best known for his theories of relativity.

Elgar, Mark A.: Australian zoologist interested in unusual fauna behaviors and animal use of chemical communication.

Emerson, Ralph Waldo (1803 – 1882): American essayist and poet.

Engelhardt, Netta: American physicist.

Euclid (~300 BCE): Greek mathematician, regarded as the father of geometry.

Exiguus, Dionysius (470 – 544): Christian monk, known for inventing the Anno Domini (AD) era.

Faraday, Michael (1791 – 1867): influential English scientist of little formal education, remembered for his contributions to understanding electrical phenomena, particularly electromagnetic induction, diamagnetism, and electrolysis.

Fermi, Enrico (1901 – 1954): Italian-born physicist, best known for his work on developing a nuclear reactor.

Feynman, Richard (1918 – 1988): eccentric American theoretical physicist who made contributions to particle physics, electrodynamics, and superfluidity.

Fields, R. Douglas: American neuroscientist.

Fodor, Jerry (1935 –): American cognitive scientist and philosopher.

Fosse, Roar: Norwegian psychologist.

Fresnel, Augustin (1788 – 1827): French engineer and physicist.

Fukuyama, Francis (1952 –): American political scientist and political economist.

Galen of Pergamon (129 – 216): Greek physician, surgeon, anatomist, and philosopher. Galen's theories influenced Western medical science for over 1,300 years.

Galileo Galilei (1564 – 1642): Italian physicist, mathematician, astronomer, and philosopher. Galileo was a seminal figure in development of science as a discipline, and a scourge to the Catholic Church for buying into Copernicus' notion of heliocentricity.

George, Henry (1839 – 1897): American political economist, journalist, and philosopher.

Geula, Changiz: American cognitive neurologist, interested in Alzheimer's disease.

Glover, Beverley J. (1972 –): English botanist.

Gödel, Kurt (1906 – 1978): Austrian logician, mathematician, and philosopher, best known for his *incompleteness theorems*, about complex systems being too difficult to prove axiomatically.

Goff, Jon: English physicist.

Golgi, Camillo (1843 – 1926): Italian physician and pathologist.

Goodman, Noah D.: American psychologist.

Gorb, Stanislav N.: Ukrainian entomologist, interested in biomechanics.

Gould, Carol G. & James L.: American ethologists and evolutionary biologists.

Gribbin, John (1946 –): English astrophysicist.

Griffiths, Thomas L.: American psychologist.

Grosseteste, Robert (1175 – 1253): English scholastic philosopher, theologian, and scientist who proposed that the universe began by expanding from a singularity of light. Grosseteste also posited the possibility of a multiverse.

Haeckel, Ernst (1834 – 1919): German biologist who conceptualized biological diversity as an evolutionary tree of life.

Harrison, George (1943 – 2001): spiritually-oriented English musician; lead guitarist of The Beatles (1960 – 1970).

Hayes, Terry: (1951 –): English author and screenwriter.

Heisenberg, Werner (1901 – 1976): German theoretical physicist, best known for asserting the *uncertainty principle* of quantum field theory, which states that measurement of subatomic particles is tricky to the point of indeterminate.

Herbert, Frank (1920 – 1986): American science fiction novelist, best known for the space opera *Dune* (1965) and its five sequels.

Hero of Alexandria (10 – 70 CE): Greek mathematician and engineer; considered the greatest experimenter of antiquity.

Hobbes, Thomas (1588 – 1679): English sociologist and political philosopher who established social contract theory and advocated despotism.

Hobson, Art: American theoretical physicist.

Holmes, Sherlock: English fictional private detective, created by Irish-Scots writer Arthur Conan Doyle.

Horava, Petr: Czech string theorist who works on D-brane theory.

Hoyle, Fred (1915 – 2001): English astronomer, mathematician, and science fiction writer. One of Hoyle's science-fiction beliefs was in a steady-state universe. Einstein shared that belief for a time.

Hubble, Edwin (1889 –1953): American astronomer, often incorrectly credited with discovery of other galaxies and galactic Doppler shift (inaptly termed *Hubble's law*). Hubble did devise the *Hubble sequence*: a simple way of classifying galaxies by how they look.

Huber, Greg: American condensed-matter physicist.

Hume, David (1711 – 1776): Scottish philosopher, historian, economist, and essayist; a logician known for empiricism and skepticism. In stark contrast to rationalists, such as Descartes, Hume believed that desire, not reason, drove human behavior.

Huxley, Thomas Henry (1825 – 1895): English biologist and anatomist, known as "Darwin's Bulldog" for his staunch advocacy of Darwinian evolution.

Jäger, Peter: German taxonomist.

Jami, Criss: American poet and philosopher.

Jenkins, Pegi Joy (1932 – 2014): American author of educational books.

Jesus (*of Nazareth*) (aka *Jesus Christ*) (7–2 BCE – 30–33 CE): Israeli Jewish carpenter and preacher who is regarded by Christians to have been the awaited Messiah (or Christ) referred to in the Old Testament. Jesus was crucified by Roman authorities for challenging societal order. (Crucifixion was reserved for crimes against the state by the lower classes, or for slaves who attacked their masters.) Though presumed literate, Jesus left no writings.

Jung, Carl (1875 – 1961): Swiss psychiatrist and psychotherapist.

Kamppinen, Matti: Finnish psychologist and philosopher of mind.

Kant, Immanuel (1724 – 1804): influential German philosopher.

Karbstein, Katrin: American molecular biologist.

Keen, Steve (1953 –): Australian economist.

Keller, Helen (1880 – 1968): American deafblind author.

Kemp, Charles: American psychologist.

Kilcher, Jewel (1974 –): American singer-songwriter, actress, author, and poet.

Kirchhoff, Gustav (1824 – 1887): German physicist who contributed to understanding electrical circuits, spectroscopy, and black-body radiation.

Kleinteich, Thomas: German biomechanist interested in vertebrate functional morphology, especially in amphibians.

Koch, Christof (1956 –): American neuroscientist, known for his work on the neural bases of consciousness.

Kollmeier, Juna A.: American astronomer.

Krystal, John H.: American psychiatrist, interested in alcoholism, schizophrenia, and post-traumatic stress disorder.

Kuhlmann, Meinard: German philosopher and physicist.

Lao Tzu (aka *Laozi*, *Lao-Tsu*, *Lao-Tze*) (6th or 5th century BCE): Legendary Chinese scholar and philosopher. His name is an honorary title. Inadvertent founder of Daoism, which teaches reverence of Nature, the value of patience, and a path to judicious existence.

It is not known when, or even whether, Lao Tzu lived. Consensus opinion among 20th century scholars is that Lao Tzu's most famous work – *Tao Te Ching* (*Daodejing*) – was a compilation by many authors. Ursula K. Le Guin notes that the work has a stylistic consistency which suggests a single primary author, with a few subsequent additions.

Lao Tzu (6th century BCE): Chinese philosopher. Inadvertent founder of Daoism, which teaches reverence of Nature, the value of patience, and a path to judicious existence.

Laplace, Pierre-Simon (1749 – 1827): French mathematician and astronomer who made important contributions to mathematical astronomy, physics, and statistics.

Le Guin, Ursula K. (1929 –): American author, best known for her fantasy and science fiction novels.

Lemaître, Georges (1894 – 1966): Belgian Roman Catholic priest and astrophysicist. Lemaître conceived the Big Bang origin of the universe, and discovered Hubble's law.

Lenin, Vladimir (1870 – 1924): Russian communist revolutionary and political theorist.

Lewandowsky, Stephan (1958 –): Australian psychologist interested in the public's understanding of science, and why people belief in falsity.

Lewis, Gilbert N. (1875 – 1946): American physical chemist, best known for his discovery of the covalent bond and his concept of electron pairs.

Lieberman, Daniel (1964 –): American paleoanthropologist.

Linnaeus, Carl (1707 – 1778): Swedish botanist, physician, and zoologist who is widely considered the father of taxonomy, despite numerous wrong guesses, including lumping amphibians and reptiles together as a single class.

Lisi, Antony Garrett (1968 –): American theoretical physicist and adventure sports enthusiast.

Locke, John (1632–1704): English philosopher and physician.

Lopez, Regis: French psychiatrist.

Luisi, Pier Luigi (1938 –): Italian chemistry professor.

Lyons, S. Kathleen: American paleobiologist.

Mach, Ernst (1838 – 1916): Austrian physicist and philosopher.

Machiavelli, Niccolò (1469 – 1527): Italian historian, politician and writer. Men who pursue their goals without regard to legal or moral limits may be called *Machiavellian*.

> Of mankind we may say in general they are fickle, hypocritical, and greedy of gain. ~ Niccolò Machiavelli

Macknik, Stephen L.: American neuroscientist.

Maharaj, Nisargadatta (born *Maruti Shivrampant Kambli*) (1897 – 1981): exceptionally insightful Indian guru, best known for the book *I Am That* (1973).

Mahler, Gustav (1792 – 1856): Austrian composer, whose exquisite symphonies often exhibit a temporal fractal quality.

Majorana, Ettore (1906 – ?): gifted Italian physicist who first predicted the neutron and Majorana fermions. His life ended mysteriously. On 27 March 1938, Majorana took a boat trip from Palermo to Naples. He disappeared; his body never found. Majorana had emptied his bank account prior to the trip. Two days before he left, Majorana wrote a note to the Director of the Naples Physics Institute, apologizing for the inconvenience that his disappearance would cause.

Makin, Simon J.: English auditory perception researcher, psychologist and science journalist.

Margulis, Lynn (1938 – 2011): American evolutionary theorist, science writer, and educator.

Markham, Edwin (1852 – 1940): American poet.

Martinez-Conde, Susana: American neuroscientist.

Mason, Malia: American psychologist.

Maupertuis, Pierre Louis (1698 – 1759): French mathematician and philosopher, who worked in classical mechanics, heredity and natural ecology. Maupertuis made the first known suggestion that all life had a common ancestor.

Maus, Gerrit: American psychologist.

Maxwell, James Clerk (1831 – 1879): Scottish physicist, most famous for formulating classical electromagnetic theory in 1865. Maxwell is widely considered the 19th century physicist most influential on 20th century physics. In 1861, Maxwell invented the first durable color photograph.

McNutt, Marcia (1952 –): American geophysicist.

McShea, Dan: American evolutionary biologist.

Medawar, Peter Brian (1915 – 1987): Brazilian-born British biologist interested in immunology.

Melcher, David: American psychologist.

Mendeleyev, Dmitry (aka *Dmitri Mendeleev*) (1834 – 1907): Russian chemist who created the modern table of periodic elements.

Michell, John (1724 – 1793): English clergyman, natural philosopher, and geologist, who made contributions in various sciences, including astronomy, geology, optics, and gravitation.

Miescher, Friedrich (1844 – 1895): Swiss physician and biologist who first identified nucleic acid.

Miesenböck, Gero: Austrian neurobiologist.

Miller, Roger (1936 – 1992): American musician, best known for the mid-1960s country/pop hits "King of the Road," "Dang Me," and "England Swings."

Mitchell, David (1969 –): English novelist.

Monaghan, Dominic (1976 –): English actor.

Moser, Jason S.: American psychologist.

Muhammad (570 – 632): Arabian religious and political leader, believed by Muslims to be the prophet of Allāh.

Mullainathan, Sendhil (1973 –): Indian economist, interested in behavioral economics.

Müller, Johannes Peter (1801 – 1858): German physiologist, most impressively known for his ability to synthesize knowledge. Müller's book *Elements of Physiology* initiated a new phase in the study of physiology, drawing from several previously distinct disciplines.

Murdoch, Iris (1919 – 1999): Irish author and philosopher.

Nagel, Thomas (1937 –): Yugoslavian-born American philosopher.

Newton, Isaac (1642 – 1727): English physicist, astronomer, alchemist, mathematician, natural philosopher, and theologian; widely considered to be one of the greatest and most influential scientists. Classical mechanics are typically termed *Newtonian physics*.

Noel, Alexis C.: American mechanical engineer.

Oppenheimer, J. Robert (1904 – 1967): American theoretical physicist, who worked alongside Enrico Fermi in developing the first nuclear weapons.

Orr, H. Allen: American biologist.

Papazian, Stefano: Swedish botanist, interested in plant physiology.

Pasteur, Louis (1822 – 1895): French chemist and microbiologist, renowned for his discoveries of the principles of vaccination, fermentation, and pasteurization. Pasteur is credited with breakthroughs in understanding the causes and prevention of diseases.

Patañjali (~250 BCE): Indian yogi.

Pauli, Wolfgang (1900 – 1958): sharp-tongued and sharp-witted Austrian theoretical physicist; a pioneer of quantum physics.

Pelli, Denis G.: American psychologist.

Penrose, Roger (1931 –): English mathematical physicist, mathematician and philosopher of science.

Petrarch (Petrarca), Francesco (1304 – 1374): Italian scholar and poet; one of the earliest humanists.

Pinker, Steven (1954 –): Canadian experimental cognitive psychologist; considered by some to be one of the world's most influential intellectuals, which is sad statement of how momentous misinformation can be.

Planck, Max (1858 – 1947): German physicist who founded quantum field theory, then rejected it out of philosophic revulsion, owing to the indeterminate nature of wave-particle duality (Heisenberg's *uncertainty principle*).

Plato (427 – 347 BCE): Greek philosopher and mathematician.

Poe, Edgar Allen (1809 – 1849): American writer.

Polchinski, Joseph (1942 –): American string theorist working on D-brane theory and interested in wormholes.

Popper, Karl (1902 – 1994): influential Austrian philosopher who rejected the classical scientific method of inductivism in favor of empirical falsification.

Proudhon, Pierre-Joseph (1809 – 1865): French philosopher who was politically a libertarian socialist.

Pusey, Matthew F.: English particle physicist.

Putnam, Hilary (1926 – 2016): American philosopher, mathematician and computer scientist.

Ramachandran, Vilayanur S. (1951 –): Indian neuroscientist, interested in vision.

Reid, Noah M.: American evolutionary geneticist.

Relman, David A.: American microbiologist and immunologist.

Revonsuo, Antti: Finnish psychologist, cognitive neuroscientist and philosopher of mind.

Roeser, Donald Brian (aka *Buck Dharma*) (1947 –): American musician, in the musical group Blue Öyster Cult (1967 – 2005).

Rosen, Nathan (1909 – 1995): American-Israeli physicist.

Rousseau, Jean-Jacques (1712 – 1778): Genevan philosopher and composer.

Rovelli, Carlo: Italian theoretical physicist.

Rudolph, Terry: English particle physicist.

Rutherford, Ernest (1871 – 1937): English physicist and chemist, known as the father of nuclear physics.

Sagan, Carl (1934 – 1996): American astronomer and science writer.

Salam, Abdus (1926 – 1996): Pakistani theoretical physicist who worked on the unification of electromagnetic and weak forces (*electroweak unification*).

Savage, Martin: American nuclear physicist.

Schaller, Mark (1962 –): American psychologist.

Schlegel, Karl Wilhelm Friedrich (1772 – 1829): German poet, philosopher, literary critic, philologist and Indologist.

Schleich, Wolfgang P. (1957 –): German theoretical physicist, interested in the foundations of quantum mechanics.

Schrödinger, Erwin (1887 – 1961): Austrian physicist and theoretical biologist who was one of the fathers of quantum field theory, and later disowned it. Best known for *Schrödinger's equation*, regarding the dynamics of quantum systems.

Schulz, Richard: American psychologist and gerontologist.

Schwarz, Norbert: German-American psychologist, interested in social psychology and consumer psychology, particularly how people form opinions and make decisions.

Schwarzschild, Karl (1873 – 1916): German physicist, best known for deriving the first exact solution to the Einstein field equations of general relativity. Einstein was only able to produce an approximate solution.

Scott, Ridley (1937 –): English filmmaker.

Seiberg, Nathan (1956 –): Israeli theoretical physicist who works on string theory.

Seifert, Colleen M.: American psychologist.

Shapiro, James A.: American molecular biologist and bacterial genetics maven.

Shaw, George Bernard (1856 – 1950): Irish playwright angered by exploitation of the working class; an ardent socialist.

Shafir, Eldar: American behavioral scientist.

Simpson, George Gaylord (1902 – 1984): American paleontologist; influential in evolutionary theory.

Smith, D. Eric: American chemical physicist.

Smith, John Maynard (1920 – 2004): English evolutionary biologist, interested in the evolution of sex.

Sorel, Georges (1847 – 1922): French philosopher.

Spinoza, Baruch (born *Benedito de Espinosa*) (1932 – 1677): Dutch rationalist philosopher who laid the philosophic foundation for the 18th century Enlightenment.

Stenger, Victor J. (1935 –): American particle physicist and godless heathen, who advocates science and reason.

Symonds, Matthew R.E.: Australian evolutionary biologist.

Sztarker, Julieta: Argentinian neuroscientist.

't Hooft, Gerard (1946 –): Dutch theoretical physicist.

Talaro, Kathleen Park: American molecular biologist.

Tamm, Igor (1895 – 1971): Russian physicist who conceptualized *phonons* in 1932.

Taroni, Andrea: English physicist.

Tegmark, Max (1967 –): Swedish-American cosmologist.

Tenenbaum, Joshua B.: American cognitive scientist.

Tertullian (Quintus Septimus Florens Tertullianus) (155 – 240): Roman Christian theologian. Unlike many Catholic church founders, Tertullian was never favored, as several of his teachings were unorthodox to later church leaders.

Thomson, Joseph John "J. J." (1856 – 1940): English physicist, credited with discovering electrons and isotopes.

Tomsic, Daniel: Argentinian neuroscientist.

Townshend, Pete (1945 –): adroit English musician who founded the musical group The Who (1964 –).

Turkheimer, Eric: American behavioral geneticist.

Turner, Michael S. (1949 –): American theoretical cosmologist and physicist.

Twain, Mark (1835 – 1910): pen name of Samuel Langhorne Clemens; talented American author, prized for his satire and wit. Best known for the novel *The Adventures of Tom Sawyer* (1876) and its sequel, *Adventures of Huckleberry Finn* (1885).

Vallortigara, Giorgio: Italian cognitive psychologist.

van Dyke, Henry (1852 – 1933): American author, educator and clergyman.

Van Raamsdonk, Mark: Canadian theoretical physicist, working on wormhole entanglement, a unified field theory.

Veneziano, Gabriele (1942 –): Italian string theorist.

Versace, Elisabetta: Italian evolutionary biologist.

von Jolly, Philipp (1809 – 1884): German physicist and mathematician.

West, Stuart A.: English microbiologist.

Weyl, Hermann (1885 – 1955): German mathematician and theoretical physicist; one of the first to conceive of combining electromagnetism with general relativity.

Wheeler, John A. (1911 – 2008): American theoretical physicist who worked on the principles behind nuclear fission. Wheeler coined the terms *black hole, wormhole,* and *quantum foam.*

Whewell, William (1794 – 1866): English polymath, scientist, science historian, economist, philosopher, theologian, and Anglican priest. Whewell's legacy is wordsmithing. He coined the terms *scientist, physicist, linguistics, consilience, catastrophism,* and *uniformism,* among others. To Michael Faraday, Whewell suggested: *ion, dielectric, anode,* and *cathode.*

Wilson, John (1943 –): English cellular biologist.

Witten, Ed (1951 –): American theoretical physicist who developed M-theory.

Wittgenstein, Ludwig (1889 – 1951): Austrian philosopher and logician, interested in mathematics, language and the mind.

> The world is the totality of facts, not of things. ~ Ludwig Wittgenstein

Wolpert, David H.: American mathematician, physicist, and computer scientist.

Yeboah, Ernest Agyemang: Ghanaian writer.

Yogananda, Paramahansa (born *Mukunda Lal Ghosh*) (1893 – 1952): Indian yogi and guru, best known for his book *Autobiography of a Yogi* (1946).

Zheng-Hui He: Chinese botanist.

Zink, Andrew G.: American behavioral ecologist.

Zwicky, Fritz (1898 – 1974): Swiss astronomer who termed *dark matter.*

Index

adaptation, 49, 60, 63, 66, 71, 116, 130
aesthetics, 128
aether, 13
Age of Enlightenment, 47
algae, 59, 67, 79
altriciality, 87
Alzheimer, Alois, 118
Alzheimer's disease, 118
Amazon molly, 127
Anaximander, 1
animals, 82
  bears, 64
  birds, 64, 88
  chimpanzees, 77
  coral, 66
  crabs, 86
  exercise, 64
  fish, 60, 62, 63
  frogs, 61
  humans, 64, 90
  insects, 71, 76
  intelligence, 82
  monkeys, 77
  rodents, 64
  tardigrades, 49
apeiron, 1
aphids, 57
archaea, 54
Archimedes, 41
Aristotle, 1, 13, 100
assumption, 103
astrocytes, 83
Atlantic silversides, 60
atomic bombs, 44
atoms, 20
Babylonia, 1
bacteria, 54
bears, 64
beauty, 128
Bell, John S., 35
Bell's theorem, 35
Big Bang, 2
biofilms, 78
birds, 88
  geese, 64
  megapodes, 88
  parrots, 88
  sunbirds, 74
black body, 16
black holes, 7, 118
bliss, 132
Bohr, Niels, 20, 22, 37
Boltzmann, Ludwig, 17
bosons, 24
braneworld, 34
Buddhism, 130
C_4, 69
Cajal, Santiago, 83
capitalism, 126
capuchins, 77
cells, 51, 124
Chandrasekhar, Subramanyan, 7
chimpanzees, 77
chlorophyll, 54
climate change, 109
coherence, 122
common ancestor, 56
consciousness, 122
Consciousness, 121
continued influence effect, 108
control, 114
convergent evolution, 68
Cooper pairs, 33
Copernicus, Nicolaus, 12
coral reefs, 66

cosmogony, 2
counterfactual definiteness, 36
counterfactual thinking, 103
crabs, 86
crystals, 28
cucumber mosaic virus, 57
curse of knowledge, 108
cyanobacteria, 54
cyclic cosmology, 3
da Vinci, Leonardo, 12
Dark Ages, 1
dark matter, 3, 5, 30, 118
Darwin, Charles, 65
D-brane theory, 34
de Broglie, Louis, 20
deception, 91
  plants, 75
decisions, 95
determinism, 101
Dirac equation, 27
Dirac, Paul, 27
dopamine, 64
dualism, 111
$E = mc^2$, 43
Earth, 50
Einstein, Albert, 2, 7, 13, 18, 32, 35
electromagnetic spectrum, 11
electromagnetism, 25
electrons, 20, 27
Elgar, Mark A., 66
emergence, 38
empathy, 91
empiricism, 50, 102
endoplasmic reticulum, 124
endosymbiosis, 59
energy, 9, 13, 117
energy conservation, 27
enlightenment, 131
entanglement, 37, 41
entropy, 10
enzymes, 52
epigenetics, 52
EPR paradox, 35
eukaryotes, 58
Euler Beta function, 33
evolution, 58, 60, 67
  saltation, 65
excitons, 32
exercise, 64
existence, 126
extra dimensionality, 15
false-consensus effect, 108
falsity, 107
fermions, 24, 27
ferns, 70
fields, 22
fish, 127
  Amazon molly, 127
  Atlantic silversides, 60
  guppies, 63
  killifish, 62
flies
  freeloader, 76
  fruit flies, 76
  gallflies, 71
flowers, 71
  color, 72
  mechanics, 73
forces, 11
framing effect, 96
freeloader flies, 76
frogs, 61
fruit flies, 76
galaxies, 3
Galen, 47
Galileo, 12
gallflies, 71
geese, 64
general relativity, 7, 14
generalization, 93
genetics, 52, 56, 68, 89, 99
ghost fields, 31
Gitterwelt, 28
glia, 83
God, 130

Gödel, Kurt, 107
goldenrods, 71
gravity, 15
Greece (ancient), 1
Grosseteste, Robert, 1
ground state, 29, 118
guppies, 63
hadron, 23
Haeckel, Ernst, 93
hartebeests, 88
Heisenberg, Werner, 21, 28, 107
Hera, 4
Heracles, 4
heuristics, 96
hierarchy problem, 31
Higgs boson, 30
Hinduism, 130
Hiroshima, 44
Horava, Petr, 34
horizontal gene transfer, 66
hornworts, 70
Hoyle, Fred, 2
humans, 64, 90
hunger, 84
illusion of knowledge, 110
immaterialism, 116, 130
immune system, 92
Impatiens frithii, 74
induction, 103
inertial reference frame, 12
influenza, 57
insects
  flies, 76
  woolly bear caterpillars, 76
intelligence, 77
  animals, 82
  microbes, 79
  plants, 79
Jupiter, 50
Kant, Immanuel, 128
kelp, 69
killifish, 62
kinematics, 23
Kirchhoff, Gustav, 16
knowledge, 100
  falsity, 107
  limits, 106
Lao Tzu, 1, 129
Laplace, Pierre, 106
Laplace, Pierre-Simon, 7
Laplace's demon, 106
lattice world, 27
laws of Nature, 104
learning, 93
Lemaître, Georges, 2
length contraction, 14
lianas, 81
life, 47, 125
light, 25
  speed, 13, 14
Linnaeus, Carl, 93
locality, 35
logic, 95
magnons, 32
Majorana fermions, 28
Majorana, Ettore, 28
Maldacena, Juan, 42
mammals
  bears, 64
  chimpanzees, 77
  hartebeests, 88
  humans, 64, 90
  monkeys, 77
  rodents, 64
  wildebeest, 88
mass, 26
materialism, 101
  philosophic, 104, 112
mathematics, 41, 104, 125
matter, 43
  dark, 3, 5
Maupertuis, Pierre, 56
Maxwell, James, 10, 13
meditation, 121, 132
megapodes, 88
Mendeleyev, Dmitry, 94

Michell, John, 7
microbes, 54, 66
  intelligence, 79
  quorum sensing, 78
microbiome, 53, 59
Milky Way, 4, 50
mind, 91, 97
mind-body problem, 112
misinformation, 107
mistakes, 113
mosses, 70
M-theory, 34
Müller, Johannes Peter, 48
muscles, 64
nattermind, 98, 131
natural philosophy, 101
neuron doctrine, 83
neurons, 83
neutrinos, 27
neutron stars, 124
Newton, Isaac, 9
nonlocality, 34, 37
noumenon, 117
orchids, 65
organs, 60
overfishing, 60
oxygen, 55
parachute plants, 76
parrots, 88
Pasteur, Louis, 48
Patañjali, 129, 130
Pauli exclusion principle, 28
Pauli, Wolfgang, 23, 27
periodic table, 94
phantom limb, 78
pheromones, 65
phonons, 32
photoelectric effect, 19
photons, 18, 25, 37, 119
photosynthesis, 68
physics, 9
placebo effect, 114
Planck constant, 17
Planck, Max, 16
plants, 65, 69
  convergent evolution, 68
  deception, 75
  decision-making, 81
  defense, 75
  defenses, 81
  electrostatic forces, 75
  ferns, 70
  flowers, 71
  goldenrods, 71
  hornworts, 70
  Impatiens frithii, 74
  intelligence, 79
  lianas, 81
  mosses, 70
  orchids, 65
  parachute, 76
  physics, 71
  prickly pear cactus, 75
  risk sensitivity, 81
  snapdragons, 72
  sundews, 80
  sunflowers, 77
  trigger plants, 74
  Venus flytraps, 80
plasmons, 32
pluralistic ignorance, 108
Polchinski, Joseph, 34
positrons, 27
precociality, 87
precocious knowledge, 87
predation, 66
prickly pear cactus, 75
proteins, 51
quanta, 17, 117
quantum
  foam, 31
  mass, 26
  mechanics, 16, 18, 34, 117
  spin, 28
  waves, 20
quarks, 23

quasars, 7
quasiparticles, 32
quorum sensing, 78
rare-male effect, 64
rationality, 95
reality, 111
reductionism, 102
relativity, 12
  general, 14
  special, 14, 26, 43
relativity of simultaneity, 14
relativity, general, 7
religion, 100
retirement homes, 115
ribosomes, 52
rodents, 64
Roger, 183
Rosen, Nathan, 42
Rutherford, Ernest, 20
saltation, 65
schizophrenia, 98
Schrödinger, Erwin, 21
Schulz, Richard, 114
Schwarzschild, Karl, 7
science, 101
scientific determinism, 106
seaweed, 69
sight, 119
sleepwalking, 120
snapdragons, 72
sociality, 66
spacetime, 4, 7, 14, 31, 41, 42, 127
special relativity, 14, 26
spin (quantum), 28
spontaneous symmetry breaking, 29
Standard Model, 22
stress, 78
string theory, 32
subconscious, 97
sunbirds, 74
sundews, 80
sunflowers, 77
supersymmetry, 31
Symonds, Matthew R.E., 66
synergy, 102, 125
Tamm, Igor, 33
Taoism, 130
tardigrades, 49
tectonics, 54
teleology, 50, 67
telomeres, 121
tensor networks, 42
theory of everything, 41
theory of mind, 91
thermodynamics, 10, 16
Thomson, "J.J.", 20
time, 38
  dilation, 14
tree of life, 93
trees, 69
trigger plants, 74
ultraviolet catastrophe, 17
uncertainty principle, 21, 107
uniqueness, 105, 126
universe, 1
  darkness, 5
  expansion, 4
  lightness, 6
vacuum, 29, 118
Van Raamsdonk, Mark, 41
Veneziano, Gabriele, 33
Venus flytraps, 80
virtual particles, 25, 30
viruses, 55
  cucumber mosaic, 57
vision, 119
vitalism, 47
von Jolly, Philipp, 16
weak force, 43
wetness, 126
Weyl fermions, 27
Weyl, Hermann, 27, 42
Wheeler, John A., 42
wildebeest, 88

willmind, 97
Witten, Edward, 33
Wolpert, David, 107
woolly bear caterpillars, 76
work, 9
wormholes, 42
Zeno effect, 40
Zeus, 4